SPACE

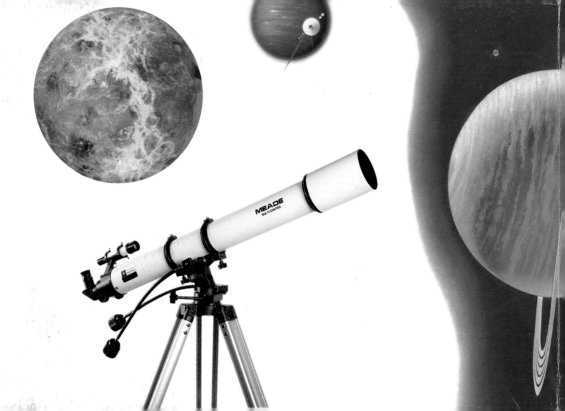

THE LITTLE GUIDES

SPACE

CONSULTANT EDITOR
Dr. John O'Byrne

FEDERAL
STREET
PRESS

This 2001 edition published by
Federal Street Press
A Division of Merriam-Webster, Incorporated
PO Box 281
Springfield, MA 01102

Federal Street Press books are available for bulk purchase
for sales promotion and premium use. For details write the
manager of special sales, Federal Street Press, PO Box 281,
Springfield, MA 01102

Publisher: Sheena Coupe
Associate Publisher: Lynn Humphries
Art Director: Sue Burk
Senior Designer: Kylie Mulquin
Editorial Coordinators: Sarah Anderson, Tracey Gibson
Production Manager: Helen Creeke
Production Assistant: Kylie Lawson
Business Manager: Emily Jahn
Vice President International Sales: Stuart Laurence

Project Editor: John Mapps
Designer: Melanie Feddersen
Consultant Editor: Dr John O'Byrne

ISBN 1-892859-25-4

Color reproduction by Colourscan Co Pte Ltd
Printed by LeeFung-Asco Printers
Printed in China

01 02 03 04 05   5 4 3 2 1

CONTENTS

PART THREE
EXPLORING SPACE

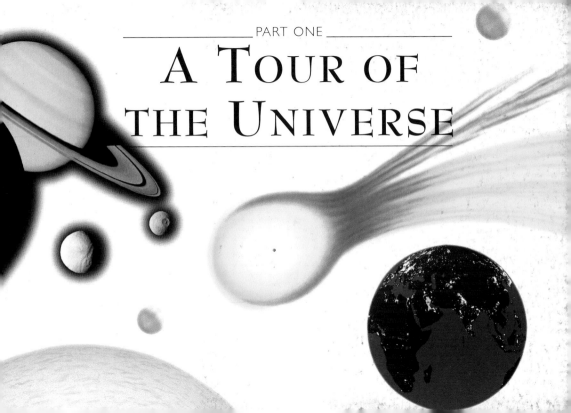

A TOUR OF THE UNIVERSE

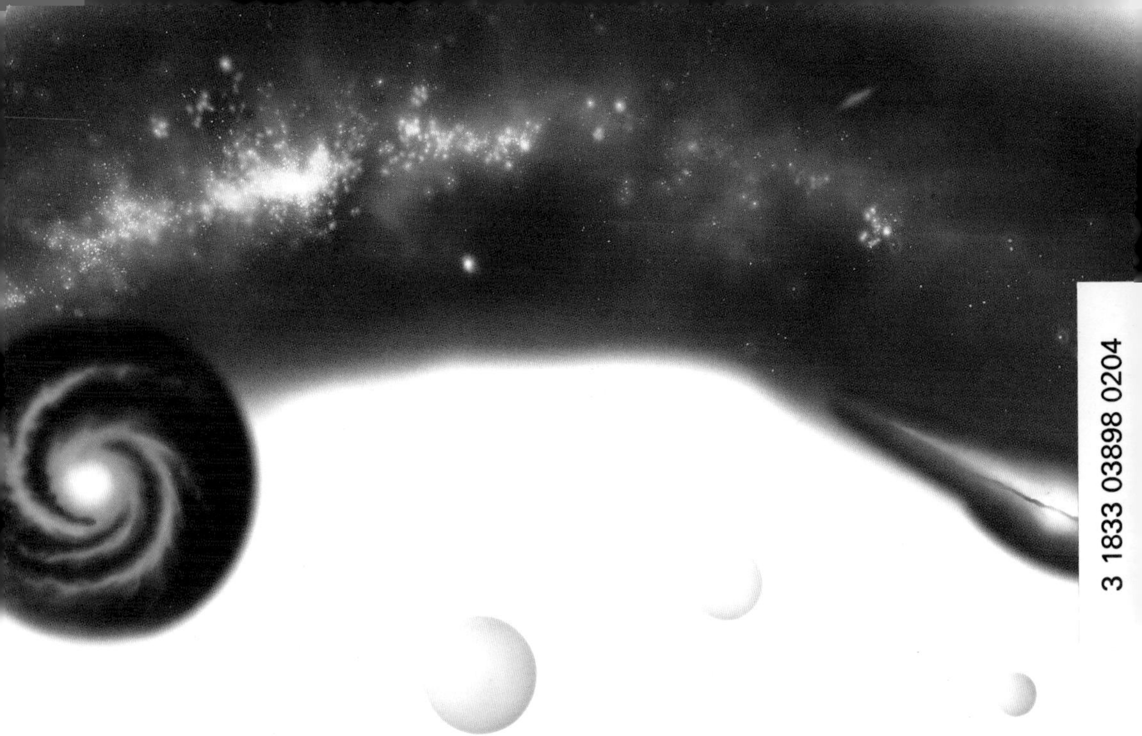

Stars
and
Galaxies

On a pitch-black night in the countryside, far from the glare of city lights, gaze upward and you will see the fascinating and beautiful spectacle of thousands of stars belonging to our galaxy, the Milky Way. Look with a telescope and you may see glittering star clusters, wispy nebulae, misty galaxies, intriguing double stars. There are boundless riches here. No wonder humans have been looking skyward for millennia, wondering how far away the stars were, what they were made of and how they got there. Today we know the answers to many of these questions, but still the fascination remains, and there are other riddles to be solved.

BEGINNINGS

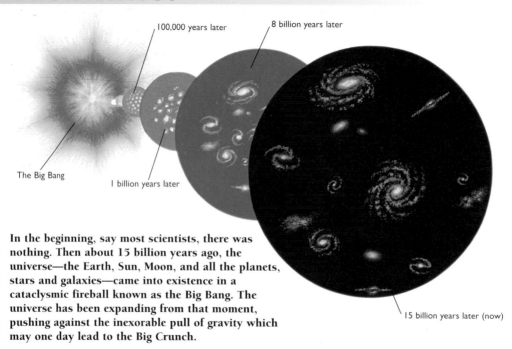

100,000 years later

8 billion years later

The Big Bang

1 billion years later

15 billion years later (now)

In the beginning, say most scientists, there was nothing. Then about 15 billion years ago, the universe—the Earth, Sun, Moon, and all the planets, stars and galaxies—came into existence in a cataclysmic fireball known as the Big Bang. The universe has been expanding from that moment, pushing against the inexorable pull of gravity which may one day lead to the Big Crunch.

MEASURING THE UNIVERSE

While a mile or kilometer is a suitable unit of distance here on Earth, when we say that the Andromeda Galaxy is 13 quintillion miles (21 quintillion km) away, the number is so large that it becomes meaningless.

Astronomical Units A useful measure in our Solar System is the Astronomical Unit (AU), the average distance between the Earth and the Sun—about 93 million miles (150 million km).

Light Years For measuring distances farther afield, we use light years—the distance light travels in a vacuum in a year. A light year is about 6 trillion miles (10 trillion km).

Parsecs Short for "parallax second," a parsec is equivalent to about 3.3 light years or 206,000 AU. It is the distance at which a star would have a parallax of 1 second of arc. Parallax is the apparent change in a nearby star's position due to the Earth's orbital motion around the Sun.

The great expansion A few billionths of a second after the Big Bang, came the first matter—minute specks of seething, exotic particles, hotter than 1,000 billion degrees. The universe cooled until more familiar particles began to form—the neutrons, electrons and protons that make up everyday matter. Gradually the particles came together to form atoms, mainly helium and hydrogen. Eventually this gas collapsed under the influence of gravity to create stars, galaxies and planets.

A dying fireball The Big Bang produced a fireball of light that has been gradually cooling ever since, and what remains from the fireball now gently bathes the universe in microwave radiation. This "cosmic microwave background," discovered by Arno Penzias and Robert Wilson in 1965, is strong evidence for the Big Bang, along with observations of galaxies which confirm that the universe is expanding.

HUBBLE'S VIEW
Each speck of light is a galaxy in this Hubble Space Telescope image. The expansion of space after the Big Bang is carrying galaxies away from each other.

The Big Crunch Gravity is slowing the universe's expansion and may cause it to stop, then go into reverse. If that happens, the universe will begin to collapse on itself, slowly at first, then speeding up until, in its last moment, all its matter will coexist as a single point.

FROM DWARFS TO SUPERGIANTS

Stars come in a truly wondrous variety. They range from dwarfs to supergiants in size, and from blue to red in color; some are intensely hot and bright, others comparatively cold and dim. How do astronomers make sense of this variety?

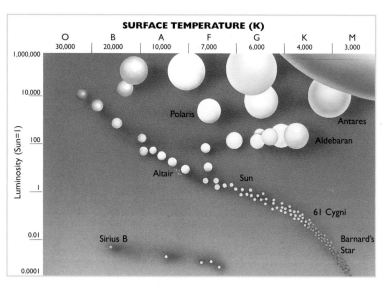

HERTZSPRUNG-RUSSELL DIAGRAM

This diagram shows the basic groups of stars, plus an indication of their color and relative sizes. Most stars lie in the main sequence, stretching from upper left to lower right.

Classifying the stars The Hertzsprung-Russell (HR) diagram classifies stars by plotting the temperature at a star's surface against its brightness, making allowance for its distance away from us. Most stars, including the Sun, fit into a diagonal band called the main sequence. These stars are often called "dwarfs," although the term is misleading: some dwarfs are 20 times larger than the Sun and 20,000 times brighter.

Red dwarfs At the cool, faint end of the main sequence are the red dwarfs, the most common stars of all. Smaller than the Sun, they are carefully doling out their fuel to extend their lives to tens of billions of years. If we could see all the red dwarfs, the sky would be thick with them, and the HR diagram would be heavy with stars in its lower right corner. But red dwarfs are so faint that we can observe only the closest ones, like Proxima Centauri, the nearest star to Earth.

White dwarfs Smaller than red dwarfs are the white dwarfs—typically the size of the Earth but the mass of the Sun. A volume of white dwarf "matter" the size of this book might have a mass of 10,000 tons! Their position at the bottom of the HR diagram marks them as different from their dwarf cousins. They are "stars" whose nuclear fires have gone out.

Red giants After the main sequence stars, the most common stars are the red giants. They have the same surface temperature as red dwarfs, but are much larger and brighter and thus lie above the main sequence in the HR diagram. These monsters typically have a mass similar to the Sun's, but, if they traded places with the Sun, their atmospheres would envelop the Solar System's inner planets.

Supergiants At the top of the HR diagram are the largest stars of all— the rare supergiants. Betelgeuse is

COMPARING SIZE
The Sun, a white dwarf, is illustrated here as a yellow globe. The red giant behind it is 100 times larger. A neutron star, thousands of times smaller, is represented by a dot.

close to 600 million miles (1,000 million km) across, which would envelop the orbit of Mars in our Solar System. Rigel, a blue supergiant, is one of the most luminous stars visible to the naked eye. One-tenth the size of Betelgeuse, it is still 100 times the size of our Sun.

THE LIFE OF A STAR

Stars begin their lives in a cloud of gaseous material, which is composed mainly of hydrogen but with a sprinkling of helium and traces of other elements too. Something happens to make the cloud unstable—a nearby supernova explosion perhaps—and the cloud begins to collapse, creating an embryo star. As the star grows, its increasing gravitational pull drags more gas inward. When gas is squeezed it becomes hotter, so as the star grows the temperature inside it rises. Eventually the star becomes so big and dense that the temperature in its core reaches 18 million°F (10 million°C)—hot enough for nuclear reactions to take place—and it starts to shine.

A STELLAR NURSERY

Deep within these columns of gas several light years in length, stars are forming. The columns are part of the Eagle Nebula (M16) in the constellation Serpens. Astronomers believe there are billions of these stellar nurseries scattered throughout the universe.

Fusion power Stars are large, gaseous balls of hydrogen and helium. A star's energy source lies at its core, where for most of its life, hydrogen is fused to form helium. Although this fusion process has been going on in our Sun for almost five billion years, it has used only a few percent of its hydrogen supplies. It is in the prime of its life—its main sequence phase.

How long to go? How long a star continues to shine depends on the amount of fuel it contains and how quickly it uses it. For example, massive blue and white stars in the main sequence have the most fuel. But they also use it quickly—the biggest stars last only a few million years before the gas runs out. A smaller star, such as the Sun, will live 10 billion years before it exhausts the hydrogen fuel in its core.

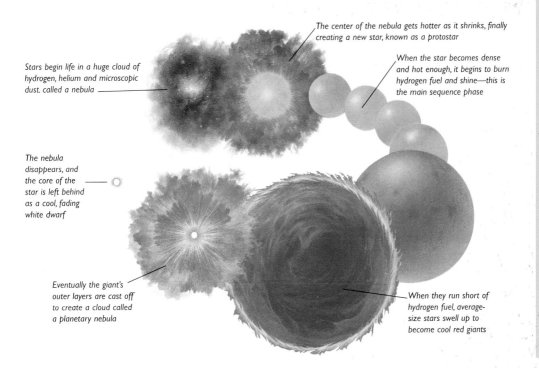

The center of the nebula gets hotter as it shrinks, finally creating a new star, known as a protostar

When the star becomes dense and hot enough, it begins to burn hydrogen fuel and shine—this is the main sequence phase

Stars begin life in a huge cloud of hydrogen, helium and microscopic dust. called a nebula

The nebula disappears, and the core of the star is left behind as a cool, fading white dwarf

Eventually the giant's outer layers are cast off to create a cloud called a planetary nebula

When they run short of hydrogen fuel, average-size stars swell up to become cool red giants

WHEN STARS DIE

Stars die when they run out of nuclear fuel, and how much fuel they have—how massive they are—determines their ultimate fate as a white dwarf, neutron star or black hole. Smaller stars, such as our Sun, end their lives relatively gently. The more massive ones go out with a bang—as supernova explosions, among the most dramatic events in the universe.

Our Sun's death When a star like the Sun starts to run low on fuel, it will grow hotter and brighter, eventually growing in size to become a red giant. After a few hundred million years at this stage, the Sun will begin to pulsate. Finally, as it exhausts its nuclear fuel, it will cast off its outer layers to form a planetary nebula surrounding a blazing hot core. This core will slowly fade away over millennia as a cooling white dwarf. Something like this is the fate of most stars.

TWO WAYS TO GO SUPERNOVA

A star at least eight times more massive than the Sun starts as a blue-white star (above). As it runs out of fuel, it swells and cools. When its fuel is spent, its core collapses and it explodes in a supernova, leaving behind a neutron star or black hole and an expanding supernova remnant of gaseous debris. In a binary system (right), a white dwarf sometimes drags material from its companion star, becomes unstable, and explodes, completely destroying the white dwarf.

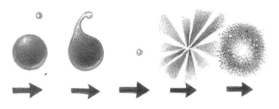

Going supernova Stars that are much more massive than the Sun reach the end of their lives more spectacularly in supernova explosions. There are two main types of supernova. The first happens when a massive star runs out of fuel; the second occurs in a binary system.

Massive stars When a star that is at least eight times the mass of the Sun runs out of hydrogen fuel, it swells up, cools down and becomes a red supergiant. It switches successively to other fuel sources, including helium, carbon and oxygen, and finally produces iron in its core. The star's furnace now shuts down. With no means of producing energy, the star collapses. The outer layers fall in on the core and then rebound, causing the explosion.

Binary supernovae In some binary systems that include a white dwarf, the companion star may "feed" stellar material into the dwarf.

If the white dwarf's gain in mass is sufficient for it to become unstable, runaway nuclear reactions take place. The end comes with an almighty explosion. Nothing is left of the original star.

Supernova remnants
However the supernova occurred, the expanding blanket of material which formed the bulk of the star collides with the surrounding interstellar medium to produce an expanding shell of gas called a supernova remnant.

Neutron stars If the collapsed core of a star survives a supernova explosion, it may remain behind in the form of a neutron star. The core becomes extremely dense and

made up entirely of neutrons—tiny particles found in atomic nuclei. A neutron star may be visible as a pulsar—a rapidly spinning "star" that flashes bursts of radio waves.

Black holes A black hole forms when an extremely massive star explodes. Too large to be a stable neutron star, the massive star keeps collapsing until it eventually disappears, leaving only a source of gravity so strong that even light waves cannot escape from it.

SUPERNOVA 1987A
A supernova was detected in 1987 in the Large Magellanic Cloud. The remnant is in the center, ringed by material ejected before the explosion.

TYPES OF STAR

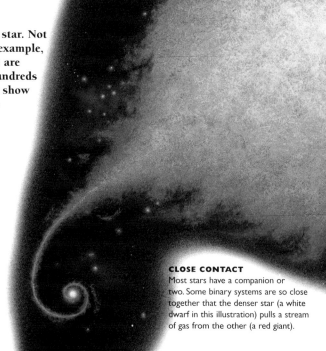

Our Sun is a stable, single, ordinary star. Not so with many other stars. Most, for example, have at least one companion. Others are grouped in clusters of as many as hundreds of thousands of members. Still more show obvious changes in brightness, some actually pulsating in size.

Stellar companions Stars in a binary or multiple system are linked together by mutual gravity and revolve around a common center of mass. Such stars probably form when several different parts of the parent cloud of dust and gas began to collapse at once. The nearest star to the Earth, Alpha Centauri, is a triplet star system. Two of the stars are bright, and orbit one another every 80 years. The third, Proxima Centauri, is a tiny red dwarf weighing only about one-tenth as much as our Sun.

CLOSE CONTACT
Most stars have a companion or two. Some binary systems are so close together that the denser star (a white dwarf in this illustration) pulls a stream of gas from the other (a red giant).

Variable stars A variable star is one whose brightness appears to change, with periods ranging from minutes to years. Mira stars are the most common type. Internal changes cause the star to pulsate in size and brightness over a period of hundreds of days. The same cause is behind Cepheid variables, which expand and contract. Still other variables are double stars that are aligned in such a way that one passes in front of the other, and then behind it, resulting in the light from the system varying—as seen from the Earth—without the stars themselves changing in brightness.

Clusters Stars may gather in clusters. Open clusters such as the Pleiades in the constellation Taurus can contain up to several hundred stars. They are loosely bound together by gravity, and perturbations from nearby stars and nebulae will shake them loose. Globular clusters are denser and more tightly bound. They contain tens of thousands of stars and have probably existed since soon after their galaxy formed.

AN ANCIENT OPEN CLUSTER
M7, an open cluster in the constellation Scorpius, contains about 80 stars. It is thought to be 200 million years old.

21

NEBULAE

Planetary nebulae are the last stage of most stars' lives

Between their stars, galaxies contain giant clouds of dust and gas which are called nebulae, the Latin name for clouds. These are the birthplaces—and sometimes the graveyards—of the stars. The nebulae we see are the thicker parts of the pervasive interstellar gas. The faintly glowing parts visible to us are lit up by light from stars within them, making them some of the most spectacular sights in the night sky.

Emission nebulae An emission nebula glows like a neon sign. It occurs when a nearby hot star shines ultraviolet light on the predominantly hydrogen gas within the nebula. The light strips electrons from the hydrogen atoms. When the electrons rejoin their parents, they give off a reddish light. Since many atoms in the nebula do this at the same time, the nebula glows red.

Reflection nebulae Nebulae of this sort shine because their dust reflects light from nearby bright stars. The small dust grains reflect blue light more efficiently than red light, so these nebulae appear blue.

Dark nebulae Dark nebulae contain the same mix of gas and dust as their bright cousins, but there are no stars nearby to illuminate them. However, we see dark nebulae when they block light coming from stars or gas behind them.

KEYHOLE NEBULA

The Keyhole Nebula is a brilliant and complex cloud of dust and gas, which is illuminated by newly formed stars. The Keyhole is part of the Eta Carinae Nebula in the constellation Carina.

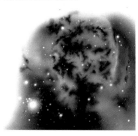

Planetary nebulae Many nebulae are the birthplaces of stars. Their dense mixtures of dust and gas—consisting mainly of hydrogen—are the raw materials for star-making. But nebulae can also signal the death of a star. After a star like our Sun evolves into a red giant, it enters a brief phase in which it blows off its outer layers. Eventually these layers become visible as a thin shell of gas around it—a planetary nebula.

HORSEHEAD NEBULA
The Horsehead is a dark nebula that can be seen only because it blots out some of the light coming from the bright nebula JC 434 behind it.

TRIFID NEBULA
The Trifid is an emission/reflection nebula in the Sagittarius constellation. It is named for the three lanes of dark clouds that divide it into three parts.

THE MILKY WAY

All of the stars visible to the naked eye are part of the Milky Way, our home galaxy. It is vast—much larger and brighter than many other galaxies in the universe. It contains roughly 200 billion stars, not to mention dense clouds of dust and gas where new stars come into existence at the rate of about 10 every year. Our own star, the Sun, is no more than a speck in one of the spiral arms, some 28,000 light years away from the galactic center.

Size and structure The Milky Way is a spiral galaxy that is flat, except at the center where there is a wide bulge. The disk of the Milky Way is about 1,500 light years thick,

with spiral arms uncoiling to a distance of 75,000 light years from the center. Surrounding this galactic disk is a halo of old stars and star clusters that stretches perhaps another 75,000 light years. Each star and nebula in this vast array orbits the galaxy's center more or less independently, our Sun completing

OUR GALAXY
The band of light across the night sky is what gave the Milky Way Galaxy its name. But in fact all of the stars visible to the naked eye are part of the Milky Way Galaxy, wherever they lie.

an orbit in about 240 million years. Every kind of star is here, from hot blue giants to cold red dwarfs.

Spiral arms Astronomers have traced relatively nearby sections of the Milky Way's spiral arms, but more distant sections are more difficult to make out, because interstellar matter gets in the way. For that reason, we are still uncertain whether the Milky Way has four spiral arms emerging from the galactic center or just two.

What is at the center? Because the Milky Way consists of so much gas and dust, it hides its secrets well, one of the greatest being what lies at its center. For some time astronomers thought that a source of strong radio emissions called Sagittarius A was at the center. Now an even smaller source of intense radiation, known as Sagittarius A*, has been found in this complex region. It might be a vast black hole with the mass of millions of suns. Material cascading into it would release the radio emissions astronomers can detect.

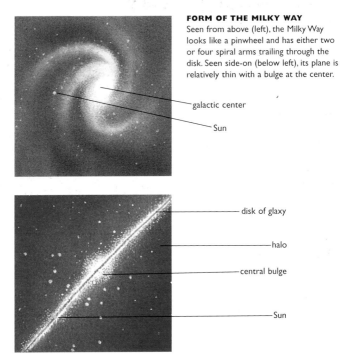

FORM OF THE MILKY WAY
Seen from above (left), the Milky Way looks like a pinwheel and has either two or four spiral arms trailing through the disk. Seen side-on (below left), its plane is relatively thin with a bulge at the center.

galactic center

Sun

disk of glaxy

halo

central bulge

Sun

GALAXIES

Stars are not randomly scattered around the universe. Instead they gather in galaxies, bound together by gravity. A single galaxy can contain hundreds of billions of stars, as well as huge clouds of dust and gas, and there are billions of galaxies in the universe. Galaxies come in a wide variety of shapes and sizes—from pinwheels, spheres and footballs to shapeless clouds.

SPIRAL GALAXY
Spiral galaxy NGC 2997 lies about 55 million light years away in the constellation Antlia.

THE LOCAL GROUP
The Local Group is a confederation of about 30 galaxies, dominated by the Milky Way and the Andromeda Galaxy. Most members are dwarf ellipticals.

Spirals Spiral galaxies range from 15,000 to 150,000 light years in diameter and may contain several hundred billion stars in a flattened disk. Within the disk, spiral arms appear to emerge from a bright central nucleus, traced out by young, hot stars and bright nebulae.

Barred spirals In barred spiral galaxies, the bright stars and ionized gas of the nucleus extend for thousands of light years from each side of the center in a straight "bar." From the end of each bar, the arms wrap back around the nucleus, as spiral arms do.

Ellipticals Elliptical galaxies tend to be shaped like footballs or spheres. They vary in size from dwarfs (with a diameter of just 1,000 light years) to giants (up to 100,000 light years across).

Irregular galaxies These are mostly faint, amorphous groups of stars, much smaller than the spiral galaxies. Approximately 5 percent of bright galaxies are irregular.

Peculiar and active galaxies Any galaxy that seems to have suffered a severe disturbance is known as a peculiar galaxy. An active galaxy has an unusually energetic core. Quasars are extremely distant objects believed to be the cores of active galaxies.

Galaxy clusters Like stars, galaxies congregate into clusters. The Milky Way belongs to the Local Group of about 30 members, but many clusters are much larger. The Virgo cluster has over 3,000 galaxies, all squeezed into a volume little greater than that of the Local Group. The Virgo cluster itself lies at the heart of a much larger collection called the Local Supercluster, which also encompasses the Local Group.

BARRED SPIRAL GALAXY
Galaxy NGC 1365 in the constellation Fornax, has particularly well-defined, scythe-like bars. It is sometimes known as the Great Barred Spiral.

THE
SOLAR
SYSTEM

In ancient times, astronomers noticed that five stars moved mysteriously through the heavens. These were not like other stars—they moved in relation to the fixed patterns of the background stars. It was the mysterious movement of these bodies that led the Greeks to assign to them their word for wanderer: *planet*. Since then, successive discoveries have revolutionized our understanding of the Solar System: the Sun and the nine major planets and their satellites, together with interplanetary material and thousands of asteroids, comets and meteoroids. And the age of discovery is not over yet—future spacecraft missions will no doubt continue to alter the picture dramatically.

29

THE BIRTH OF THE SOLAR SYSTEM

The Sun and the Solar System were born in violence and chaos. Some 4.6 billion years ago, a large cloud of cold dust and gas was drifting around the center of the Milky Way. The cloud began to collapse, possibly set off by the explosion of a nearby star. Eventually a forerunner to our Sun was born at its center. What happened next? The generally accepted theory—shown on these pages—is that grains of material from the cloud consolidated into solid lumps of material, which grew into larger bodies, which in turn became the planets we see today.

1. A WHIRLING CLOUD

The cloud of dusty gas from which the Solar System was born begins to rotate as it collapses, perhaps after a nudge from a supernova explosion. Gravity and orbital motion force it to settle into a relatively thin disk rotating around the proto-Sun, all spinning in the same direction.

2. PLANETISIMALS TAKE SHAPE

Lumps of dust in the disk collapse into solid bodies called planetisimals. Near the proto-Sun these are rocky in composition, while the planetisimals farther away were cool enough to retain water, gases and ice.

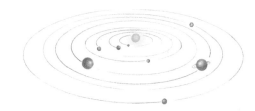

4. PROTOPLANETS GROW

With protoplanets nearly at full size,
gravitational encounters and collisions
among them can tilt their rotation axes—
or eject them from the Solar System.

5. A FINAL "CLEAN-UP"

Finally, after no more than 100
million years, the newborn Sun
abruptly brightens, and its
radiation blows away any material
that has not been swept up by
the planets—and the formation of
the Solar System is complete.

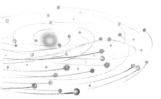

3. VIOLENT COLLISIONS

Planetisimals collide and gradually form
larger bodies termed protoplanets. As
these grow by sweeping up the
planetisimals in their vicinity, collisions
become more violent.

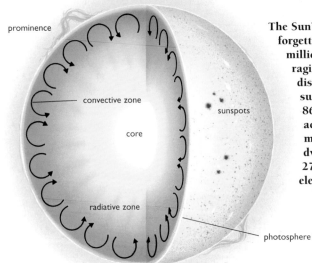

- prominence
- convective zone
- sunspots
- core
- radiative zone
- photosphere

NUCLEAR POWERHOUSE
The Sun's photosphere—the visible surface—underlies boiling convection layers. Deeper down, energy streams out from the nuclear powerhouse at the core.

The Sun's very ordinariness tricks us into forgetting that we live less than 100 million miles (160 million km) from a raging thermonuclear reactor. That disk drifting across the sky each sunny day is a ball of seething gases 865,000 miles (1.4 million km) across that is about 333,000 times more massive than the Earth. This dwarf star is 71 percent hydrogen, 27 percent helium, and other elements add up to just 2 percent.

A nuclear furnace The Sun's gaseous core has a temperature of 27 million°F (15 million°C). In the core, hydrogen atoms bang into each other constantly. In a series of reactions, four atoms can fuse into one atom of helium and release a tiny amount of energy. These energy sparks, released in huge numbers,

SUN FACT FILE

Distance from Earth 93 million miles (150 million km)
Sidereal revolution period 365.26 days
Mass (Earth = 1) 333,000
Radius at equator (Earth = 1) 109
Sidereal rotation period (at equator) 25.4 days
Apparent size 32 arc minutes

power the Sun and thereby give us life. Every second, the Sun fuses about 600 million tons of hydrogen.

Moving heat Each tiny energy parcel spends millions of years being absorbed and re-emitted before reaching the surface. Fusion occurs from the center of the Sun out to perhaps a quarter of its radius. Above this core lies a region—the radiative zone—in which radiation carries the energy. This is where the energy spends most of its time, trickling outward.

Convective zone On top of this layer the Sun's energy moves like boiling water. Heated from below, the gas rises to the surface, radiates energy into space, cools, then sinks again. This convective region forms the outer third of the Sun. Deep inside, the bubbles of hot gas are huge, but at the surface that we on Earth can see—

the photosphere—they break down into granules that are about 500 miles (800 km) across.

Photosphere and beyond The photosphere is where sunspots form. These are small areas where the solar magnetic field stops hot gas from reaching the surface. Cooler than their surroundings, sunspots appear dark. Their numbers rise and fall as part of a poorly understood 11-year cycle. Above its visible surface, the Sun has a complex atmosphere which consists of the chromosphere (a thin, cool layer) and the corona, which is almost as hot as the core. Both are usually only visible during solar eclipses. Beyond the corona, a "solar wind" of protons and electrons blows into space.

SOLAR PROMINENCES

This image highlights a gigantic solar prominence. Prominences are areas of cooler hydrogen gas caught in the high-temperature corona.

THE PLANETS

A family of nine planets and their satellites orbits the Sun. As spacecraft continue to explore the Solar System, scientists are learning that each planet is a world of its own, differing greatly from the others in almost every way. Yet we can sort this family into two broad groups based on size, density and chemical makeup—the terrestrial planets and the gas giants.

Terrestrial planets Mercury, Venus, Earth and Mars are the terrestrial planets. These are small bodies made of rock, with densities three to five times that of water. Terrestrial planets have comparatively thin atmospheres, and Mercury has no atmosphere to speak of at all.

Sun

Mercury

Venus

Earth

Mars

Jupiter

Saturn

Uranus

Neptune

Pluto

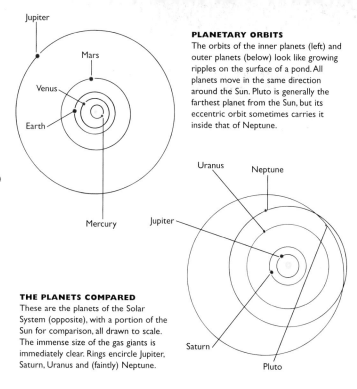

PLANETARY ORBITS
The orbits of the inner planets (left) and outer planets (below) look like growing ripples on the surface of a pond. All planets move in the same direction around the Sun. Pluto is generally the farthest planet from the Sun, but its eccentric orbit sometimes carries it inside that of Neptune.

THE PLANETS COMPARED
These are the planets of the Solar System (opposite), with a portion of the Sun for comparison, all drawn to scale. The immense size of the gas giants is immediately clear. Rings encircle Jupiter, Saturn, Uranus and (faintly) Neptune.

Gas giants Jupiter, Saturn, Uranus and Neptune are all more than a dozen times more massive than the Earth. While each has a tiny rocky core, it is buried under layers of hydrogen and helium thousands of miles deep. Each of these gas giant planets has a density near that of water, and Saturn is actually less dense than ice. Pluto does not fall into either gas giant or terrestrial category. It is more like an icy moon.

More than just planets Between Mars and Jupiter lies the asteroid belt, a zone of rocky debris stopped from coalescing into a planet by the gravity of Jupiter. Meteorites are fragments of asteroids, swept up by the Earth after they were broken off their parent bodies by impacts. Every planet, except Mercury and Venus, has at least one moon, and each of the four gas giants has a ring system. Finally, on the fringes of the Sun's domain, halfway to the nearest stars, lies the realm of the comets.

MERCURY AND VENUS

Two of our closest neighbors, Mercury and Venus are very different from our own congenial world. Both are extraordinarily barren and inhospitable—Mercury with its extremes of heat and cold, and Venus with its sulfuric acid clouds and scorching temperatures.

MERCURY FACT FILE
Distance from the Sun 0.39 AU
Sidereal revolution period (about Sun) 88.0 days
Mass (Earth = I) 0.055
Radius at equator (Earth = I) 0.38
Sidereal rotation period (at equator) 58.7 days
Moons None
Apparent size 5–13 arc seconds

VENUS
Light-colored mountain belts scrawl across Venus's surface in this radar image from the Magellan mission.

MAPPING VENUS
Magellan's aim was to map the surface of Venus using a radar altimeter. This image shows a mountainous terrain.

Mercury's extremes Mercury has virtually no atmosphere, so there is no protection to shield its surface from the intense barrage of solar radiation it receives and to smooth out the variation in temperature between day and night. Its sunny-side temperature is as much as 750°F (400°C), and then it drops to minus 330°F (–200°C) at night.

Pock-marked world Mercury has a heavily cratered surface. While the planet generally resembles the Moon, it does not have the dark "seas" of lava found on the lunar surface. Instead, its gently rolling lava plains appear more like the lunar highlands in composition, although they are much less heavily cratered. Beneath the surface, Mercury has an unusually large nickel-iron core, which fills three-quarters of the planet's radius. It may have resulted from a collision early in the planet's history that removed part of Mercury's mantle.

Venus's hell Venus is shrouded by thick clouds of water vapor and sulfuric acid, which form in an atmosphere that is almost entirely carbon dioxide. This atmosphere has produced a runaway greenhouse effect—the Sun's radiation heats the surface but the heat cannot escape into space. Temperatures on the surface approach 860°F (460°C)—the hottest in the Solar System— and midnight is as hot as midday.

Surface features The Pioneer and Magellan probes discovered a planet-wide museum of features—buckled and folded mountain ranges, enormous volcanoes, extensive lava flows, sinuous lava channels and chunky lava domes. The rarity of impact craters suggests that the surface is relatively young, perhaps 500 million years old. Scientists think that the planet must have experienced a global volcanic cataclysm that abruptly erased the traces of earlier eras.

VENUS FACT FILE
Distance from the Sun 0.72 AU
Sidereal revolution period (about Sun) 225 days
Mass (Earth = 1) 0.81
Radius at equator (Earth = 1) 0.95
Sidereal rotation period (at equator) 243 days
Moons none
Apparent size 10–64 arc seconds

LAVA FLOW
Sif Mons, one of Venus's many volcanoes, is 1.2 miles (2 km) high. A lava flow (in yellow) extends for hundreds of miles from the peak.

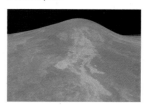

EARTH

Among all the planets, the Earth is unique in several respects. To an observer elsewhere in the Solar System, it displays a dynamic atmosphere that is remarkable in having a 21 percent oxygen content. Clouds of water vapor obscure a variable portion of the planet's surface, but make it a brilliant beacon in the inner Solar System. Perhaps most remarkable of all is the fact that over 70 percent of its surface is covered in water. Liquid water is not seen anywhere else in the Solar System.

Oceans cover almost three-quarters of the Earth's surface

Inside the Earth Internally, the Earth consists of core, mantle and crust. The nickel-iron core is partly molten, and is about 4,400 miles (7,000 km) in diameter. The mantle extends from the top of the core almost to the surface. It contains high-density basaltic rocks, rich in iron and magnesium. The mantle's top layers lie beneath the crust—where we live. The thickness of the crust varies, being 5 or 6 miles (10 km) thick in the ocean basins, but reaching a depth of 50 or 60 miles (100 km) under the continents. The crust is broken into a dozen or so tectonic plates that collide and interact, riding on the back of the warm, plastic rock in the upper mantle.

Atmosphere Above the Earth's surface lies the atmosphere, an ocean of air rich in nitrogen and oxygen. It sustains life, protects us from the Sun's higher energy radiation, and drives our weather (by interacting with the Sun's heat).

Watery world Water is found in oceans (which cover 70 percent of the Earth's surface), lakes and rivers; below the surface as groundwater; locked up in frozen polar ice caps and glaciers; and as vapor carried in the atmosphere. The Earth is the only planet where temperatures allow surface water to exist in solid, liquid and gaseous states.

EARTH FACT FILE
Distance from the Sun 1.00 AU
Sidereal revolution period (about Sun) 365.26 days
Mass (Earth = 1) 1.0
Radius at equator (Earth = 1) 1.0
Sidereal rotation period (at equator) 23.9 hours
Moons 1

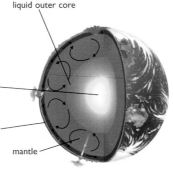

liquid outer core

solid inner core

crust

mantle

DOWN TO THE CORE
Radioactivity in the Earth's inner and outer cores generates heat, which creates warm currents in the mantle.

PROTECTIVE BLANKET
The Earth's atmosphere is a blanket of air that keeps the planet comfortably warm. It is so thin that if the planet were the size of an apple, the atmosphere would be only as thick as the peel.

THE MOON

The Earth is unique among the inner planets in having a large natural satellite—the Moon. The favored theory as to how this came about is that about 4.5 billion years ago a huge object struck the Earth, melting the object plus most of the Earth and sending a spray of rock into space. The spray cooled into a ring of rocky debris, which eventually coalesced into the Moon.

The face of the Moon has appeared essentially unchanged for more than half Earth's history

Craters and seas For 500 million years after its formation, the Moon was bombarded by asteroids and meteorites. The biggest impacts created wide basins, hundreds of miles across, that subsequently became flooded with lava. These dark regions are known as maria (singular, mare) or "seas." The remainder of the Moon's surface, unaffected by lava flooding, forms the bright and intensely cratered highlands. Craters range from tiny pits to huge walled plains. Relatively recent impacts are responsible for the bright streaks called rays.

CAPTURED
A collision early in the Earth's history produced a cloud of rocky debris that orbited the Earth. The debris formed clumps which slowly collected together. The result was a new body that cooled down to become the Moon.

LUNAR ECLIPSE
When the Earth passes between the Moon and the Sun, a lunar eclipse occurs for everyone on the nighttime side of the Earth. During a total eclipse, the entire Moon is in the shadow of the Earth.

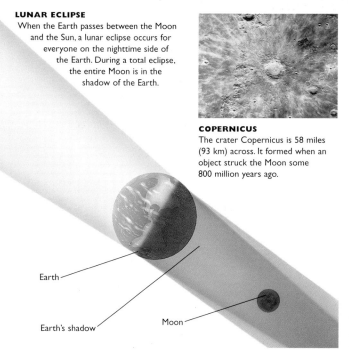

COPERNICUS
The crater Copernicus is 58 miles (93 km) across. It formed when an object struck the Moon some 800 million years ago.

Earth

Earth's shadow

Moon

MOON FACT FILE
Distance from the Earth
239,000 miles (384,000 km)
Sidereal revolution period (about Earth) 27.3 days
Mass (Earth = 1) 0.012
Radius at equator (Earth = 1) 0.272
Apparent size 31 arc minutes
Sidereal rotation period (at equator) 27.3 days

Phases As seen from the Earth, the Moon passes through a series of phases every 27.3 days—waxing from new Moon, through first quarter, to full Moon, then waning to last quarter and new Moon again.

Tides The Moon's gravity, and to a lesser extent that of the Sun, causes two high tides on the Earth each day. One bulge of water, on the side nearest the Moon, marks where gravity is pulling the water away from Earth. The other bulge, on the opposite side, marks where the Moon is pulling Earth away from the water.

41

MARS

With clouds, storms and seasons, Mars is the most Earth-like of the Sun's family. For all its small size, being roughly half as big as the Earth (its diameter measures 4,217 miles, or 6,787 km), the planet Mars has certainly loomed large in our imaginations. This little red world draws our attention (and our spacecraft) as few others do.

MARS FACT FILE
Distance from the Sun 1.52 AU
Sidereal revolution period (about Sun) 687 days
Mass (Earth = 1) 0.11
Radius at equator (Earth = 1) 0.53
Sidereal rotation period (at equator) 24.6 days
Moons 2
Apparent size 4–25 arc seconds

MARS FROM VIKING
In 1976, two Viking landers probed the surface of Mars. One of the accompanying orbiters sent back this image of the Martian disk.

A brutal climate Mars has a 25 degree tilt, giving the planet four distinct seasons, although its climatic extremes exceed anything we experience on Earth. The atmosphere—95 percent carbon dioxide—is thin and offers only a small barrier to escaping heat. The surface temperatures barely reach 32°F (0°C) by day and drop to minus 190°F (−123°C) at night.

In the arctic regions in winter, the atmosphere directly deposits dry ice.

Red desert The Martian world is a desert one, and the planet's ruddy color (detectable even by the naked eye) comes from its rusty, oxidized rocks and dust. A telescope shows an ocher-colored surface with darker markings. Once believed to be areas of vegetation, these are now known to be vast lava flows and boulder fields. Windblown sheets of fine dust and sand cause the changes in these markings that misled early observers into thinking they had seen evidence of plants or lichens.

Water Mars has channels hundreds of miles long, down which enormous floods of water have raced, scarring the terrain and carving out stream-lined islands. Where is the water now? A little is locked in thin polar ice caps of water and carbon dioxide ices. Perhaps the rest lies in permafrost beneath the surface.

Surface features Mars has some spectacular features, one of the most prominent being Olympus Mons, an enormous volcano, seemingly dormant, which is larger than any mountain on the Earth. The Valles Marineris is equally remarkable—a system of canyons up to 4 miles

MAKING TRACKS
Sojourner, the rover carried by Mars Pathfinder, makes its mark on the dusty surface of Mars in 1997. Here it uses its spectrometer to gather data about the rock in the center of the photograph.

(7 km) deep, forming a great gash stretching 2,500 miles (4,000 km) across the planet.

JUPITER

The ancients chose Jupiter's name more wisely than they knew. This bright planet, which takes 12 years to circle the sky, has been associated in many mythologies with the most powerful of the deities. Perhaps it was the symbolism of spending a year in each constellation of the zodiac that was particularly impressive. In any case, Jupiter also ranks as the king of the planets on other grounds. It has a diameter of 89,400 miles (143,800 km) and has more mass than all the other planets put together.

JUPITER
This color-enhanced image from Voyager I shows tremendous details. The colors and banding result from chemical reactions in the ammonia and methane in the Jovian atmosphere.

A mini-Sun Jupiter's makeup is remarkably like that of the Sun—hydrogen accounts for 75 percent and helium 24 percent; the rest is mainly methane and ammonia. There are other parallels with the Sun: Jupiter's immense gravity directs the fate of many comets, and it can send asteroids careening through the Solar System. It also governs a miniature Solar System of 16 moons.

probably somewhat similar to a big terrestrial planet—a molten ball of silicate rock that is several times the mass of the Earth.

In a spin Jupiter's rapid rotation—less than 10 hours—smears its cloud features into east–west stripes paralleling the equator. The rapid rotation also causes Jupiter's marked oval shape—the planet is 7 percent wider at its equator than at its poles.

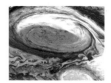

GREAT RED SPOT
The Great Red Spot is a vast high-pressure system, covering an area about twice the size of the Earth. We do not know its origins, but it tends to become redder as solar activity increases.

Beneath the clouds The "surface" of Jupiter consists of layers of cloud standing near the top of an immense atmosphere thousands of miles deep. Below the top of the cloud deck, the atmosphere remains gaseous to a depth of perhaps a few hundred miles. Beneath this, high pressures and temperatures turn hydrogen into a deep layer of molecular liquid, which is followed by a layer of metallic liquid hydrogen. Finally, Jupiter's core is

GAS GIANT
Jupiter's atmosphere consists almost entirely of hydrogen and helium. At the planet's center lies a core of molten rock.

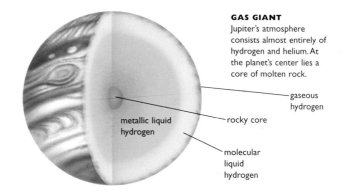

gaseous hydrogen

rocky core

metallic liquid hydrogen

molecular liquid hydrogen

THE MOONS OF JUPITER

When Galileo made the first telescopic observations of Jupiter in January 1610, he spotted four pinpoints of light accompanying the planet. He had discovered Jupiter's four largest moons. The planet has 12 other satellites, but the Galilean moons remain the focus of most interest. They are among the most fascinating bodies in the Solar System.

Galilean moons Jupiter's four largest moons, from Jupiter outward, are Io, Europa, Ganymede and Callisto.

Io Io is 2,256 miles (3,630 km) in diameter. It is the most volcanically active body in the Solar System. As it circles Jupiter, it is tugged by the gravity of Jupiter, Europa and Ganymede. This tidal flexing keeps Io's interior molten, and causes repeated eruptions, which obliterate signs of impact craters.

LITTLE SOLAR SYSTEM
This composite Voyager image shows Jupiter and its Galilean moons. Callisto (right) and Ganymede (lower left) are both mixtures of ice and rock; Europa (center) and Io (far left) have greater proportions of rock.

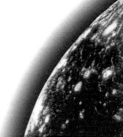

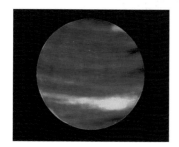

IO

Io is constantly remaking its surface through volcanic action. Detailed features change in a matter of months.

Europa Europa is 1,950 miles (3,138 km) in diameter. The smooth, bright crust is scarred with a network of darker lines; craters are few, implying the surface is young. Scientists believe that Europa has a rocky core and a deep ocean of water or slush, covered by the icy skin seen by spacecraft. The dark lines resemble freshly frozen openings in the polar ice pack on the Earth.

Ganymede Ganymede is the largest Jovian moon—and the largest moon in the Solar System—at 3,270 miles (5,262 km) in diameter. It contains a mixture of rock and ice. Its face, rich in water-ice, has many craters, yet it also has regions of younger, grooved terrain. Scientists are not sure what caused the grooves but an expansion of Ganymede's

EUROPA

The surface of Europa is almost pure water-ice. Does it cover a water ocean which might harbor life?

interior as it melted throughout could have generated these volcanic/tectonic features. The activity apparently continued for some time until it froze in place.

Callisto Callisto is 2,983 miles (4,800 km) in diameter. It is the least altered of the Galilean moons, and is densely cratered. The moon appears generally dark with silicate "dirt," but most of the craters have bright interiors, ray patterns and exposed fresh ice. An impact basin 900 miles (1,500 km) across shows where an asteroid or comet struck.

Lesser moons The remaining 12 worlds within the Jovian system tend to be small and heavily cratered. The ones that are closest to Jupiter are the most rocky, while those on the fringes have more ice in their make-up. The outermost four moons revolve around Jupiter in a back-ward direction and are probably captured asteroids.

SATURN

Saturn was the last of the planets known to antiquity. As it crept around the sky, taking nearly 30 years to complete a circuit, it seemed to embody the infirmities of old age. So the planet was labeled Saturn—the father of the gods in Greek and Roman mythology. After the invention of the telescope, Saturn, with its exquisite rings, became the showpiece of the Solar System.

FLOATING WORLD
The density of Saturn is low, less than that of water, in fact. In principle, the planet would float.

Bands of clouds Saturn has a smaller diameter (75,000 miles or 120,660 km) than its fellow gas giant, Jupiter, and it is considerably lighter, with a mass equal to 95 Earths. Its composition also parallels that of Jupiter: 74 percent hydrogen, 24 percent helium, and small amounts of methane, ethane and ammonia. Chemical reactions by the latter three cause Saturn's tan color and faint banding. Like Jupiter, Saturn is flattened at the poles, though its degree of flattening is greater. The surface we see is crossed by cloud bands.

Inside Saturn Scientists think that Saturn's inner structure resembles that of Jupiter. A layer of clouds covers a thick layer of fluid hydrogen that grows hotter and denser the farther it is from the surface. This probably becomes metallic about 20,000 miles (30,000 km) down. The core is thought to be a molten silicate ball weighing a dozen or more Earth masses.

SATURN FACT FILE
Distance from the Sun 9.54 AU
Sidereal revolution period (about Sun) 29.5 years
Mass (Earth = 1) 95.2
Radius at equator (Earth = 1) 9.5
Sidereal rotation period (at equator) 10.2 hours
Moons 18
Apparent size (planet's disk) 15–21 arc seconds

At the cloudtops Since Saturn orbits farther from the Sun than Jupiter, its environment is colder. This means it has less "weather" and so displays fewer features in its cloudtops.

The rings While Jupiter, Neptune and Uranus also have ring systems, none of them is as grand or as easy to see as Saturn's. The rings of Saturn span 170,000 miles (270,000 km) and tilt 29 degrees to the planet's orbit. They are no more than a few hundred yards thick, but can be seen in backyard telescopes.

Icy chunks The rings are made up of innumerable icy chunks that range in size from fine dust to house-size blocks. They form the three broad rings visible from the Earth, which are lettered from the outermost A, B and C. The Voyager craft recorded hundreds of very narrow "ringlets" making up the rings. Between the bright A and B rings lies the Cassini Division, 2,600 miles (4,200 km) wide, which appears to be a gap, but was revealed by Voyager to be merely less densely packed with ringlets. While ring particles are mostly ice, they contain rocky impurities that alter their colors.

How the rings formed The rings are probably no more than 10 to 100 million years old, which is short-lived

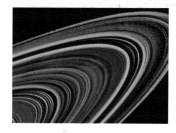

THE RINGS FROM A TO C
This image spans Saturn's rings from the inner C ring (bluish color at left), through the B ring (warm tints merging into greens and blues), out to the Cassini Division (dark with bright streaks). Outside the Cassini Division is the A ring.

in Solar System terms. They exist because one or more moons collided or came close enough to Saturn for its gravity to break them apart. Scientists think that in a few tens of millions of years the rings will disappear as collisions between particles slow their orbits and the remains fall into Saturn.

THE MOONS OF SATURN

SATURN'S SATELLITES
This composite image of pictures taken by Voyager 1 shows Saturn surrounded by four of its 18 moons. Dione is at the front with Tethys to its right and Mimas, with its gigantic crater, at far right. To the left of Saturn's rings lies Enceladus.

Saturn has an extended family of satellites—18 at last count. They consist primarily of rock and water-ice with craters dominating the surface of most of them. They range in size from tiny Pan, at 12 miles (20 km) across, up to Titan, which is 3,200 miles (5,150 km) in diameter.

Titan Bigger than Mercury, Titan is the second largest moon in the Solar System, after Jupiter's Ganymede. Its atmosphere—it is the only satellite known to have one—is 50 percent thicker than the Earth's. While the atmosphere is primarily transparent nitrogen, it also contains methane, ethane, acetylene and other compounds. Exposed to sunlight, these produce a smog which hides the moon's surface features. Scientists

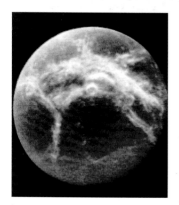

DIONE
We know little about Dione. The first close-range images of it, including this one, were taken by Voyager 1 in 1980.

think that at Titan's temperature (minus 292°F or −180°C), pools of liquid methane may exist.

Iapetus Saturn's oddest moon may well be Iapetus, which is 907 miles (1,460 km) across. Like our Moon and most other satellites, Iapetus keeps one side turned toward the planet it orbits. What makes this moon unusual is that the forward-facing hemisphere is jet-black in color, while the one that is trailing is snow-white. The white material appears to be ice, but the nature of the jet-black material remains a mystery. It may be dust that has been chipped off Phoebe (the next moon outward) and swept up by Iapetus—or it could have erupted through some internal process.

Hyperion Hyperion is too small a moon for its gravity to overcome its irregular shape: it measures 255 by 162 by 137 miles (410 by 260 by 220 km). Heavily cratered, Hyperion does not orbit with its longest axis aimed toward Saturn, which would be its natural state. Planetary scientists think that an asteroid or comet may have hit the moon recently enough that it has not yet settled back into alignment.

Enceladus Enceladus appears to be the most geologically active of Saturn's moons, with the activity being driven by heat produced inside Enceladus by repeated gravitational tugs from its neighbor moon Dione. Enceladus has a diameter of 300 miles (500 km) and a density indicating that it consists largely of water-ice. It displays a bright surface with a mixture of old, well-cratered areas and newer terrain that is grooved and fissured.

Mimas The moon Mimas is fairly small, being 242 miles (390 km) in diameter. The brightness of its surface and its low density are indications that it has an icy composition. Unlike Enceladus, however, it appears to be old and heavily cratered, which is a common fate in the Solar System. Among its craters is a particularly large one named Herschel, the result of a collision that must have almost torn the moon apart.

URANUS

The English astronomer William Herschel stumbled upon Uranus one evening while he was surveying stars in the constellation Gemini. The date was March 13, 1781, and Uranus thus became the first planet to be found in modern times. This distant world remained poorly understood until recent exploration by Voyager 2.

Blue-green world Uranus has a diameter of some 31,765 miles (51,120 km), nearly four times the size of the Earth. It weighs as much as 14 Earths, and mimics the Sun in composition: hydrogen and helium comprise almost all of Uranus. Traces of methane in the atmosphere give Uranus its soft blue-green color.

Side-on Uranus orbits the Sun tipped almost on its side. This probably resulted from a collision early in its history.

URANUS
In this artist's conception, Uranus is shown surrounded by its narrow, thin rings and three of its 15 moons.

URANUS FACT FILE
Distance from the Sun 19.2 AU
Sidereal revolution period (about Sun) 84.0 years
Mass (Earth = 1) 14.6
Radius at equator (Earth = 1) 4.0
Sidereal rotation period (at equator) 17.9 hours
Moons 15
Apparent size 3–4 arc seconds

HOW URANUS GOT ITS TILT
Most of the planets in the Solar System are tilted (the Earth, for example, has a tilt of 23 degrees), but Uranus is tilted almost on its side. One theory to explain the tilt says that early in its history Uranus collided with an object the size of the Earth (left).

AFTER THE CRASH
The force of the crash was great enough to produce the current tilt of 98 degrees (below). The planet's rings and moons may be leftover fragments from the collision.

UMBRIEL'S FACE
Umbriel is a heavily cratered moon, and shows no sign of geological activity.

Uranian rings Narrow and thin, Uranus's rings are kept in place by the gravitational effects of tiny moonlets. A typical ring particle lasts perhaps 500 years before being destroyed when it collides with other particles or spirals into Uranus.

The moons Uranus has 15 known moons, the most remarkable being Miranda. It has a 6-mile (10 km) cliff at the edge of a fault zone, and at least three huge areas with roughly concentric grooves.

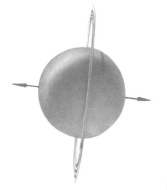

NEPTUNE

This planet is named after the Roman god of water, in earlier times also a god of fertility. This relationship with the sea has turned out to be highly appropriate, as photographs of Neptune show it to be a deep, crisp blue. The Voyager 2 spacecraft, which flew past in 1989, took images which showed bright clouds of methane ice crystals floating in the atmosphere.

VANISHING SPOT
The Great Dark Spot—a giant storm—was photographed by Voyager 2 in 1989. Five years later, images from the Hubble Space Telescope showed that the spot had disappeared.

NEPTUNE
The Voyager 2 probe showed Neptune as a big, blue ball, with many markings and cloud bands. There is also a faint ring system, not visible here.

A planet of water Neptune is believed to have a small, rocky core, but the bulk of the planet is probably a deep ocean of water. This then merges into an atmosphere of hydrogen and helium. Methane in the top of the atmosphere gives the planet its pronounced blue color.

NEPTUNE FACT FILE
Distance from the Sun 30.0 AU
**Sidereal revolution period
(about Sun)** 165 years
Mass (Earth = 1) 17.1
Radius at equator (Earth = 1)
3.88
**Sidereal rotation period (at
equator)** 19.2 hours
Moons 8
Apparent size 2.5 arc seconds

disappearance of some features, notably the Great Dark Spot, and the appearance of new ones.

Triton Neptune has a retinue of eight moons. The largest, Triton, is 1,680 miles (2,706 km) in diameter, about two-thirds the size of our Moon. It is a mixture of ice and water, covered with pinkish frosts of nitrogen and methane condensed from the atmosphere. It has geysers that shoot plumes of nitrogen gas 5 miles (8 km) into the sky. Winds then blow the gas plumes about 60 miles (100 km) downrange. Triton has plains and impact craters, as well as a strange cantaloupe area of pits and depressions crossed by ridges.

Clouds and spots Voyager 2 saw atmospheric bands, bright ammonia-ice clouds, and the Great Dark Spot—a storm the size of the Earth. Winds in the equatorial zone fly westward at 900 miles (1,500 km) per hour. Neptune radiates more than twice as much heat as sunlight provides and this is thought to drive much of the atmospheric activity. Since the Voyager visit, the Hubble Space Telescope has recorded the

ERUPTIONS ON TRITON
Triton, Neptune's largest moon, is volcanically active. Its volcanoes erupt nitrogen, not hot lava.

Captured comet? Triton revolves around Neptune in a backward direction, the only large moon in the Solar System to do so. Planetary scientists think it was captured by Neptune, maybe after drifting in from the Kuiper Belt, a region of planetesimals—comets without tails—that starts roughly at Neptune and extends outward a thousand astronomical units or more. However, it has a much rockier composition than the planetesimals. Perhaps it changed during capture, which possibly involved collisions with other, now-vanished, moons.

PLUTO

In February 1930, astronomer Clyde Tombaugh discovered the last of the Solar System's planets. It was named Pluto for the Roman god of the underworld. Tombaugh had found an odd world, a planet we still know relatively little about.

A COLD WORLD
The shadow of the moon Charon falls on the icy surface of Pluto. Charon is about half the size of Pluto. The distant Sun casts a cold light on these far-off worlds.

Rock and ice Pluto has a rock and ice core covered with layers of ices. Its surface may resemble Neptune's moon Triton, with nitrogen and methane frosts. The temperature is about minus 370°F (–220°C). The evaporating frosts give it a thin atmosphere of nitrogen and methane, which its feeble gravity is slowly losing to space. But as Pluto's orbit takes it farther from the Sun, the atmosphere will recondense and fall as frost.

PLUTO FACT FILE
Distance from the Sun 39.5 AU
Sidereal revolution period (about Sun) 249 years
Mass (Earth = 1) 0.002
Radius at equator (Earth = 1) 0.18
Sidereal rotation period (at equator) 6.39 hours
Moons 1
Apparent size 0.04 arc seconds

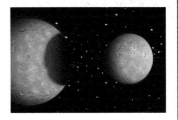

PLANET AND MOON
With only 12,100 miles (19,500 km) separating them, Pluto and Charon are so close that they are almost a "double planet." Charon, discovered in 1978, orbits the planet every 6.4 days.

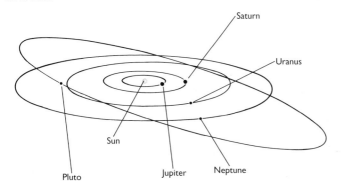

Saturn

Uranus

Sun

Pluto

Jupiter

Neptune

PLUTO'S ORBIT
Relative to the orbits of all the other planets, Pluto's is more tilted and eccentric. At times this brings the planet closer than Neptune to the Sun. It takes Pluto 249 years to complete one revolution around the Sun.

ASTEROIDS

In 1801, the Italian astronomer Giuseppe Piazzi discovered the first of the minor planets. He named this small, rocky object Ceres. The starlike appearance of Ceres lent it the alternative name these bodies are still known by: "asteroid." More discoveries followed, and today astronomers have identified 35,000 asteroids. Most of these orbit the Sun in a belt between Mars and Jupiter; others, however, have orbits that approach the Earth's. Wherever they lie, they promise to tell us much about the development of our Solar System.

IDA
Asteroid Ida is 32 miles (52 km) long, with a heavily cratered surface.

Battered remnants Asteroids are the remains of a much larger population of small bodies that formed and evolved under the gravitational control of Jupiter. Collisions have steadily reduced larger asteroids to smaller ones, and smaller ones to dust and fragments.

ASTEROID ORBITS
Most asteroids orbit the Sun in a belt between Mars and Jupiter. More asteroids are located in Jupiter's orbit, while others (not shown) range more widely.

The asteroid belt Most asteroids orbit within the main belt, the region between Mars and Jupiter, and their orbital periods last several years. Those closer to the Sun have compositions that are more purely rocky, while those nearer Jupiter contain additional organic and carbon-related compounds. A few asteroids have surfaces as bright as weathered concrete, but most are dark, rather like asphalt pavement.

Trojans Not all asteroids orbit within the main belt. Collisions among them, and tugs from Jupiter, have sent asteroids on courses that cruise the Solar System. Jupiter has two clutches of asteroids traveling with it in its orbit, one ahead and one behind. They are known as the Trojan asteroids.

Near-Earth asteroids Other asteroids have trajectories that approach the Earth or even cross its orbit, with some risk of collision. For example, in 1993 a small asteroid approached to within 90,000 miles (140,000 km) of the Earth—well within the Moon's orbit. Many scientists believe that the dinosaurs were wiped out in the aftermath of a small asteroid striking the Earth 65 million years ago, and that nothing but luck stands in the way of another impact occurring at any time.

TOUTATIS
The heavily cratered asteroid Toutatis actually consists of two bodies, and is known as a binary asteroid.

We know of 200 asteroids that have orbits which cross the Earth's orbit, and there are probably 10 times that number we do not know about.

Keys to the past Astronomers are interested in asteroids because these objects provide a view into the distant past. They preserve relatively intact one part of the solar nebula from which the planets formed. Some asteroids are primitive shards of rock, little altered from the moment they cooled 4.6 billion years ago. Others are fragments of larger bodies that may have undergone some geological evolution.

Finding names When an astronomer discovers an asteroid, he or she earns the right to suggest a name. Early discoveries bear names from classical mythology, but antiquity has been ransacked by now. So today names include 6000 United Nations, 2985 Shakespeare and 4659 Roddenberry (of *Star Trek*).

COMETS

Comets are masters of making something out of very little. When a bright comet such as Hyakutake or Hale-Bopp pays us a visit, we delight in its ghostly luminance and huge tail, which may be millions of miles long. Few of us are aware that all the glowing magnificence is provided by the comet's nucleus—a "dirty snowball" no bigger than a small city.

Snowballs in space When a comet is far from the Sun, it is a cold body, perhaps a few miles across. Most reside in a vast sphere of comets surrounding the Sun—the Oort Cloud—beyond the orbit of the most distant planets. A comet is occasionally perturbed onto a path toward the Sun. As it closes in, ice begins to boil away, and a coma of gas develops. The material leaves the comet to form separate gas and dust tails streaming away from the Sun.

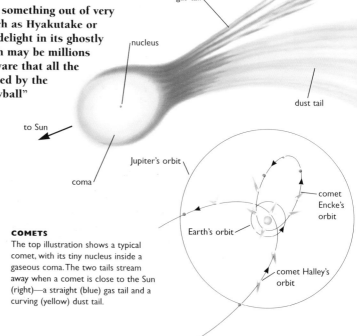

gas tail

nucleus

dust tail

to Sun

coma

Jupiter's orbit

Earth's orbit

comet Encke's orbit

comet Halley's orbit

COMETS
The top illustration shows a typical comet, with its tiny nucleus inside a gaseous coma. The two tails stream away when a comet is close to the Sun (right)—a straight (blue) gas tail and a curving (yellow) dust tail.

Periodic comets Sometimes a comet will pass close to a planet, usually Jupiter, and the planet's gravity changes the comet's orbit. Repeated encounters may result in a new orbit that causes the comet to return repeatedly to the inner Solar System. Periodic comets that return to the vicinity of the Sun within 200 years are known as short-period comets. The most famous of these is Halley's, which returns every 76 years. With a period of 3.3 years, Encke's comet has the shortest period. Comets Hyakutake and Hale-Bopp are examples of long-period comets. Comet Hyakutake passed close to the Earth in March 1996, its first return in 9,000 years; comet Hale-Bopp returned a year later, after 4,000 years.

The origins of life? Comets hitting the Earth may have provided the building blocks for life on our planet—carbon, hydrogen, oxygen and nitrogen. Observations of Halley's comet in 1986 determined the presence of these materials in almost identical amounts as exist on the Earth. It is possible that comets also brought our water supply. Astronomers have calculated that comet Hale-Bopp, the great comet of 1997, carried one trillion tons of water, an amount 50 percent greater than all the waters in North America's Great Lakes.

HALLEY'S COMET
Recorded sightings of Halley's comet go back over 2,000 years. It last visited the Earth's skies in 1986, when this photograph was taken. It will return in 2061.

METEORS AND METEORITES

IMPACT SITE
Arizona's Meteor Crater is about 3,600 feet across and 600 feet deep (1,100 by 180 m). It was formed 50,000 years ago when an iron meteorite about 150 feet (45 m) in diameter struck with a force of 15 megatons of TNT.

On any clear, dark night, even in the city, you probably can spot a meteor—a "shooting star"—if you watch the skies for a while. This brief flash of light was once interpreted as a dead person's soul on its way to heaven, or even as a warning of death. The scientific explanation, however, is that a piece of space grit known as a meteoroid is burning up in the Earth's upper atmosphere.

Meteoroids A meteoroid enters the atmosphere at a speed ranging from about 6 to 45 miles (10 to 70 km) per second. Friction with air molecules heats and vaporizes it, resulting in the bright streak we call a meteor. Some meteors leave faintly glowing trains—trails of ionized gas—behind them.

Comet debris According to scientists, most meteors are debris shed by comets, which are a mixture of ices and dust. When a comet nears the Sun, its ices evaporate, freeing the dust, which slowly spreads out from the comet's orbit.

Showers At certain times of year, the Earth sweeps through the trail of dust from a comet, and a meteor shower occurs. Meteor showers take their name from the constellation the meteors appear to come from, and several dozen showers recur every year. For example, the meteor shower that appears to emanate from the constellation Gemini every December 14 is called the Geminids, meaning "children of Gemini." The Perseids of August are another well-known shower. Meteors that do not appear to belong to any shower are called sporadics. Most of them probably belonged to one shower or another long ago but the showers are now so depleted as to be unrecognizable.

MAJOR ANNUAL METEOR SHOWERS			
Shower	**Date**	**Hourly**	**Parent comet**
Quadrantids	January 3	40	
Lyrids	April 22	15	Comet Thatcher
Eta Aquarids	May 5	20	Comet Halley
Delta Aquarids	July 28	20	
Perseids	August 12	50	Comet Swift-Tuttle
Orionids	October 22	25	Comet Halley
Taurids	November 3	15	Comet Encke
Leonids	November 17	15	Comet Temple-Tuttle
Geminids	December 14	50	Asteroid 3200 Phaethon
Ursids	December 23	20	Comet Tuttle

Dates can vary slightly. The hourly rate represents the number of meteors you might see under a dark sky when the radiant—the point in the sky from which the meteors appear to radiate—is near the zenith. Expect to see perhaps half as many more if the shower is strong.

Meteorites Most meteors burn out high in the atmosphere and their minute residue slowly drifts down to the ground. But not all interplanetary material arrives so gently. Larger fragments survive the passage through the atmosphere to impact on the ground. A fragment found intact is called a meteorite. They display many different compositions and weigh from a few dozen ounces to many tons.

PART TWO

OBSERVING THE SKY

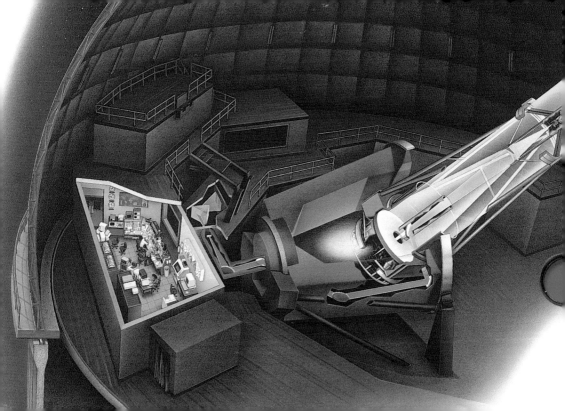

TOOLS
AND
TECHNIQUES

Astronomy is an activity for a lifetime of learning and wonder. You may start out by watching a beautiful grouping of the Moon and planets and noting the progress of stars and constellations through the seasons. A pair of binoculars, which you may already own, opens up new vistas—from the Milky Way resolved into thousands of stars, to a glowing comet or the shuttling moons of Jupiter. A telescope takes you into deep space to explore such sights as star clusters, colorful nebulae and spiral galaxies, as well as spots on the Sun. This chapter gives some practical tips on viewing and advice on optical instruments, so that your skywatching can become an even richer experience.

HOW TO BEGIN OBSERVING

Exploring the heavens—by eye, binocular or telescope—lets you explore the universe firsthand. You may never travel in space, but anyone can use a telescope as a spaceship of the imagination. All it takes on your part is a willingness to step out of the daily routine, a desire to be enthralled and, most importantly, a spirit of adventure.

CAPTIVATING CHILDREN
For children, observing the night sky is a great spur to the imagination. And they don't need binoculars or a telescope to make this an enjoyable pastime.

LOOK UPWARD AND HAVE FUN
Young or not-so-young, novice astronomers should learn key stars and constellations before buying binoculars or a telescope. The emphasis at first, especially for children, should be on having fun—they could try connecting the stars to make their own patterns.

Starting out Many people buy a telescope far too soon. If you do not yet own a telescope, do some naked-eye observing until you know your way around the sky. Take every opportunity—at an observatory or club—to look through a variety of telescopes. When you are ready to buy one of your own, see the guide-

lines in this chapter. If you already own one, the advice in this chapter will help you make the most of it.

The sky in motion Skywatching with the naked eye will not only make you familiar with the main constellations but will accustom you to the changing sky. The Moon, for example, rises about 40 minutes later, on average, every night, so its position relative to the stars differs each time it rises. The planets trace much slower paths among the stars, and the stars themselves rise about four minutes earlier each night.

Recording the sky You will find keeping a sky diary or logbook very satisfying. It can be simply a list of what you see each night, embellished with some descriptions. You could also add sketches.

Choosing a site Your observing site will have a greater effect on your skywatching than any piece of

A SKYWATCHING KIT
Some skywatchers spend a fortune on elaborate equipment. The kit shown here, for example, includes everything from a hot drink in a flask to a computerized tracking system. High expenditure is not necessary, however; your own curiosity will largely determine how far you go.

equipment. Of course, a dark site far from city lights is ideal, but few of us have that luxury. City-bound observers may not be able to find faint galaxies, but the Moon and planets can still look splendid. For the dedicated, a short drive beyond the city limits is often enough to get out from under the dome of light that covers every city and town.

Join a club After skywatching on your own for a while, you might find it useful to join a group of like-minded people. Most major cities have astronomy clubs. Joining a club almost guarantees that you will have all the help you need to get your telescope working properly. If you are thinking of buying a telescope, you could attend a club viewing night and try the various types that members own. Should you be interested in serious observing, you may find people who will help you get started with observing variables or patrolling for novae outbursts.

BINOCULARS

Binoculars will allow you to see the craters on the Moon, four moons of Jupiter, and five or 10 times as many stars as can be seen with the naked eye. Dozens of asteroids come within reach of binoculars each year, while bright comets look spectacular. Not bad for a piece of equipment costing only about a quarter of the price of a standard beginner's telescope.

PORRO PRISM
The shape of porro-prism binoculars results from the prism arrangement which produces an image that is upright and the correct way round. Roof prism types use a different arrangement of prisms to achieve the same result.

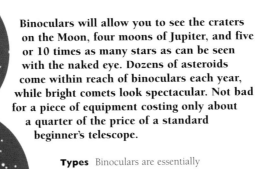

COMPARE THESE VIEWS
Most 7x binoculars show about 7 degrees of sky (top). Higher-power 10x models magnify the image more, but usually show only 5 degrees of sky (above).

Types Binoculars are essentially two low-powered telescopes joined together, so that you can look through them with both eyes instead of just one. The two main types of binoculars differ mainly in the orientation of the prisms that are used. The porro-prism design is simpler than the roof prism, and is the one most commonly used in binoculars today. The roof prism is more expensive, but is quite a bit lighter and more compact.

Size The performance of binoculars depends on the diameter of the objective lenses (the lenses at the front) as well as on the magnification that is provided by the eyepieces. Various combinations of lens and eyepieces provide a considerable range of possibilities. For night viewing, experienced observers recommend using a pair of 7 x 50 binoculars, which always use the porro-prism design. The 7 refers to the magnification and the 50 refers to the diameter of each of the objective lenses (in millimeters). While 10 x 50 or 8 x 40 binoculars are also suitable for using in the dark, they do not match the light that is available to the dark-adapted eye quite so well. Binoculars with 80 mm objective lenses are also commonly available, but a pair that large should have a mounting.

Where to buy Many cheap binoculars sold by department stores are useless, for unless the two optical paths are exactly aligned, you will see slightly different images with each eye. As your eyes struggle to compensate, viewing becomes uncomfortable. It is best to buy from a dealer of optical equipment.

TRIPOD
Your binoculars have to be held steady to get the best views. You can simply lean against a wall or fence, or ideally, mount the binoculars on a sturdy camera tripod.

CHILD-FRIENDLY
Binoculars are easy to use, which is especially important for kids.

TELESCOPES

A telescope can show you details on the Moon as tiny as a mile across, the stunning rings of Saturn, the changing clouds of Jupiter, and the brightest galaxies, nebulae and star clusters of deep space.

What to look for Size and stability should be your main considerations when buying a telescope. As a rule of thumb, buy the telescope with the largest mirror or objective lens you can afford. Usually a reflecting telescope will provide larger optics than a refractor of the same size. The result is more light coming into your eye and brighter images of faint stars and galaxies. Also make sure that the telescope is well mounted: a wobbly stand will make your observing sessions difficult and ineffectual.

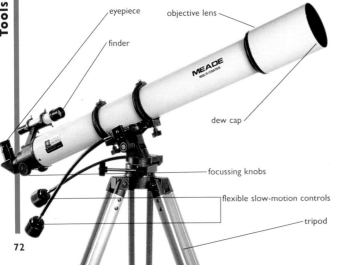

eyepiece

objective lens

finder

MEADE
MULTI-COATED

dew cap

focussing knobs

flexible slow-motion controls

tripod

TELESCOPE SAVVY
This 90 mm refracting telescope sits on a good-quality altazimuth mount. Look for a telescope with all metal and wood construction, with minimal use of plastic, especially on any moving parts.

REFRACTING TELESCOPE
This type uses a lens to gather light, which is focused to an eyepiece at the end that magnifies the image.

REFLECTING TELESCOPE
Relectors use a mirror to collect light, which is reflected and focused back up the tube to the eyepiece

CATADIOPTRIC TELESCOPE
These telescopes use a mirror and a corrector lens. The light travels back and forth then exits through a hole in the mirror to the eyepiece.

TYPES OF MOUNTING
The altazimuth mount is widely used on small refractors (below left). The equatorial mount (below right) permits motorized tracking of the stars.

Other factors Refractors tend to give sharper, brighter images than reflectors of the same aperture because their optical elements are far less likely to slip out of alignment. However, refractors of about 4 inch (100 mm) aperture are physically far bigger than their reflecting cousins—an important consideration if you want to move your telescope around.

Some options If you have a limited budget, you might initially opt for a reflector with a 4 inch (100 mm) diameter mirror, whereas

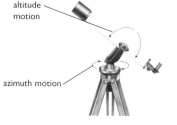

altitude motion

azimuth motion

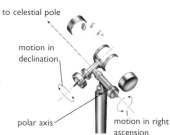

to celestial pole

motion in declination

polar axis

motion in right ascension

a more serious (or wealthy) skywatcher might choose one of 8 inch (200 mm) diameter or more. Buying a good-quality 2.4 or 3 inch (60 or 75 mm) refractor might also be a wise choice for a beginner.

Where to buy Go to a reputable dealer rather than a department store. Prices may be a little higher, but you will find amateur astronomers on staff who can help you make a wise selection.

73

ACCESSORIES

Whenever you look through a telescope, you look through an eyepiece. Eyepieces are usually the first accessories telescope owners buy. A few other well-chosen accessories, including solar filters, motor drives and computer controls, can improve the performance of your telescope and increase your enjoyment of stargazing.

GOING WIDE

The Great Nebula in Orion looks the same size through a 20 mm Kellner (far right) and a 20 mm wide-angle eyepiece (near right), but the wide-angle shows you more of the sky.

KEEPING OPTICS CLEAN

To clean an eyepiece, first blow off loose dust with a lens blower. Then lightly moisten a cotton swab with camera-lens cleaning fluid. Wipe the lens gently with the wet swab, then again with a dry one to remove any streaks.

Eyepieces Eyepieces are used with a telescope to provide magnification. A wide variety is available, giving an extensive range of magnification and fields of view. Quite often telescopes are sold with only one inexpensive eyepiece and you will have to buy additional ones.

Focal length Eyepiece focal lengths are measured in millimeters. The lower the focal length, the higher the magnification. Beginners will find a 25 mm eyepiece most useful, plus perhaps a 12 mm one. On a typical 2.4 inch (60 mm) refractor, these eyepieces would yield magnifications of 28x (28 times) and 60x respectively. The same eyepieces on a typical 4 inch (100 mm) reflector would yield 40x and 85x. The full Moon would not quite fit in the field of view at 85x.

Finder A finder lets you search through a wide area of sky to center the telescope on an object. It provides a wide field of view (about 5 degrees) at only 5x to 10x magnification. The Telrad, a recently developed alternative, appears to project a faint red bulls-eye toward the sky, which you then center on the object you are searching for.

Solar filters Many small telescopes are sold with solar filters that fit in or near the eyepiece. Never use one of these as it can cause eye damage. If you

A RANGE OF ACCESSORIES
Eyepieces (back) and colored eyepiece filters. The Barlow lens (lower left) is used with an eyepiece to double its magnification. The camera adaptor (right) couples a camera to a telescope.

want to use a solar filter, buy a large one that can be attached to the front of the telescope.

Motor drives A motor drive enables your telescope to follow a star as it moves across the sky. If you have a drive on an equatorial mounting or a computer-controlled altazimuth mounting, you can watch an object without having to touch the telescope at all. If you use an equatorial mount, the telescope's polar axis must be pointed accurately to the north or south celestial pole. A drive is essential for taking long-exposure photographs of planets and deep-sky objects.

Computer controls Most of the more expensive telescopes can now be controlled through a computer. You simply key in a series of instructions and the telescope will then find whatever you wish.

SETTING UP YOUR TELESCOPE

Make the best possible start with your new hobby by reading this step-by-step guide to setting up your telescope. Then for the first few weeks, concentrate on the objects that you can see with the naked eye, particularly the Moon and the bright planets. By far the easiest object to find, the Moon rewards you with a different show each night. You will soon find yourself travelling along the crater rims, climbing the lunar mountains and exploring the valleys.

Advice from an expert when purchasing can make setting up easier.

FINDING THINGS

A star chart helps you to locate faint objects. Converting dots on a page into stars in the sky is a process that takes some getting used to, so do it slowly. First find a small group of bright stars on the chart, then find them in the sky. Proceed star by star, from the map to the sky, until you find the spot that contains the object you are looking for.

Using the finder, center on a star that is close to that object—preferably within 1 degree (or a fingertip held at arm's length). The finder usually has a field of view of 6 degrees. Now use the main telescope and the lowest power eyepiece to move the telescope slowly in the direction of the object you are seeking. Finding it might take a few tries.

Step 1 Line up the finder with the telescope. This is best done during the day by viewing a distant treetop, or at night by looking at a street light. The adjustment can be done on a star, but it will be more difficult because of the star's motion across the sky.

Step 2 If your telescope has an altazimuth mount you can simply take it outside, put it down in any position and start observing. If it has an equatorial mounting, the polar axis of the mounting must be pointed toward the north or south celestial pole. This is easily done in the Northern Hemisphere by finding Polaris. Sigma (s) Octantis, the 5th magnitude southern pole star, is harder to find. To make it easier to find the pole, make sure that the mounting is level; that the polar axis points upward from the horizontal at the angle of latitude of your site; and that the polar axis points north–south (true north and not magnetic north).

Step 3 Insert the eyepiece with the longest focal length. This will give you the lowest magnification and the widest field of view, and thus the best chance of finding things.

Step 4 Choose a bright object like the Moon, a bright planet, or a star. Center the object in the finder, and then look through the main telescope. If the finder is not perfectly aligned with the main telescope you will need to pan the telescope very slowly—up, down and across. This might take time, but eventually you will find the object (and be convinced that aligning the finder carefully is worth the trouble). Once this happens, lock the mounting's axes. This moment is called first light—the moment the telescope first sees light from an object in the sky.

Step 5 The star will probably look like a blob of light to begin with, or maybe like a doughnut if you have a reflecting telescope, which will mean the eyepiece needs focussing. Slowly move the focussing knob in one direction; if the blob or the dough-nut grows bigger, move it the other way until the star focusses into a point. Success!

GETTING ALIGNED
For an equatorial mount to work properly, it must be polar aligned to the celestial pole. In the Northern Hemisphere, this means aiming the polar axis at Polaris, the northern pole star. For the Southern Hemisphere, the pole star is the faint Sigma Octantis.

motion of star

north celestial pole

S

E

N

OBSERVING THE SUN SAFELY

The Sun is a source of endless fascination. A solar eclipse is one of the most awe-inspiring events you can see in the skies. And watching, drawing or photographing sunspots as they change is an intriguing area of stargazing to complement your nighttime observations. But studying the Sun has its hazards—looking directly at its disk can cause permanent damage to your eyes. Fortunately, a few simple precautions will ensure safe viewing.

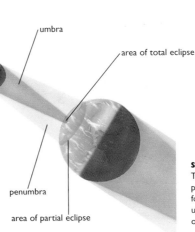

SOLAR ECLIPSE
Viewed from the Earth, the Moon and the Sun look about the same size. Thus the Moon can eclipse the Sun by passing in front of it. The inner, darkest part of the Moon's shadow is the umbra, the outer shadow, the penumbra.

SAFE VIEWING
This telescope is equipped with a Sun protection screen. The Sun's image is focussed onto the lower screen, while the upper screen shields the image from other sunlight.

Sun warning Never look through a telescope directly at the Sun without proper protection. Even a split second of unfiltered light could permanently blind you. Solar eye-piece filters are worthless; throw

TOTALITY

You can watch the progress of a solar eclipse through a telescope rigged for solar observation. At the moment of totality, when the Moon has fully covered the Sun—as in this photograph—you can look directly with the naked eye. But remember to turn away as soon as the Sun reappears.

that is coming through the eyepiece onto a piece of paper, shielding this "screen" from direct sunlight so that you can see the Sun's image clearly. Focus the eyepiece until the image is sharp. Cover the telescope aperture when it is not in use to prevent heat build-up in the telescope.

Filters An alternative to the projection method is to use a full-aperture solar filter well secured to the *front* of your telescope. The best types are made from reflective mylar or glass.

FILTERS
The large solar filter lets you safely observe a solar eclipse. The smaller, colored filters are for viewing planets.

them away if they came with your telescope. Never use the finder to find the Sun either, since looking at the Sun through an unfiltered finder is also extremely dangerous.

Solar safety To observe the Sun safely, this is what you should do.

Look at the way the shadow of the telescope is falling on the ground and move the telescope around until its shadow is at a minimum. The Sun should then be shining down through the telescope and out through the eyepiece, toward the ground. Project the image of the Sun

UNDERSTANDING
THE
SKY

If you are a newcomer to astronomy, the night sky can seem a bewildering place. This is, after all, unfamiliar territory where nothing carries a signpost, and it's easy to feel lost. However, the sky's many marvels and graceful cycles and motions are well within your grasp. This chapter is intended to enhance your astronomical knowledge by giving you a solid grounding in some of the basics—about how astronomical coordinates work; about the sky and how and why it appears to move; about stars and their different brightnesses and colors; about time and the seasons; about the planets and their movements; and about the best ways to find your way around the night sky.

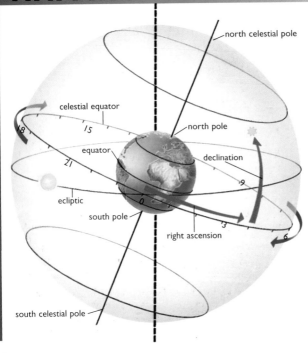

The sky extends infinitely far, filled with stars of all types, brightnesses and distances. But to our eyes, it appears as a hemisphere covering us like a dome. We can adopt this impression, and call it the celestial sphere to include that part hidden by the Earth underfoot. The sphere allows us to locate objects in the sky with coordinates similar to latitude and longitude.

CELESTIAL SPHERE

The imaginary celestial sphere encircles the Earth, appearing to turn westward once a day, as indicated by the blue arrow. Of course, it is the Earth that really spins daily on its axis. That axis is tilted relative to the Earth's orbit, causing the Sun's path among the stars—the ecliptic—to meet the celestial equator at an angle.

Sky latitude As latitude measures distance north or south of the Earth's equator, *declination* measures angular distance from the celestial equator. It runs from 0 degrees at the equator to 90 degrees (north and south) at the poles, and is measured in degrees, minutes and seconds of arc. North of the celestial equator declination is listed as positive (+), while south is negative (−).

Sky longitude The celestial equivalent of longitude is called *right ascension*, or RA, and is measured in hours, minutes and seconds (where 24 hours equals 360 degrees and one hour of RA equals 15 degrees of arc). It is measured eastward from the point where the Sun's path crosses the celestial equator from north to south.

The altazimuth system This enables astronomers to specify the position of a celestial body with respect to their horizon and at a particular time using coordinates

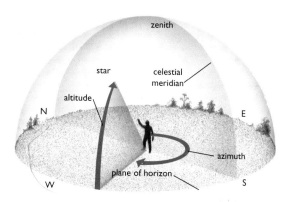

called altitude and azimuth. These figures differ for each observer, depending on his or her position on Earth and the time that the observation is made. *Altitude* (also known as elevation) is the angle above the observer's horizon; the point directly overhead, at 90 degrees, is known as the *zenith*. *Azimuth* is the angle measured clockwise from north along the horizon to the point on the

ALTAZIMUTH COORDINATES
The altazimuth system is especially useful to explain where an object is in the sky at any particular moment. The meridian is an imaginary line that runs due north and south and passes through your zenith.

horizon that lies beneath the star (see illustration above). Thus, N = 0 degrees or 360 degrees azimuth; E = 90 degrees; S = 180 degrees; and W = 270 degrees.

A SPINNING EARTH

Our planet Earth is a roughly spherical globe which revolves around its own axis, an imaginary line that runs through the center of the Earth from the north geographic pole to the south geographic pole. This spinning motion and our location on the globe are two of the major factors that determine what we see in the night sky at any given time.

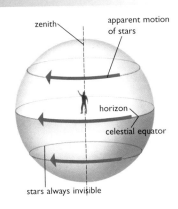

AT THE NORTH POLE, 90°N
Only stars in the northern half of the celestial sphere are visible.

A shifting horizon As the Earth rotates on its axis from west to east, our perception is that the stars, the Sun and all the other heavenly objects move around us in the opposite direction—that is, from east to west. The apparent motion of the stars in the sky, however, will depend on where you happen to be on Earth.

The role of latitude The concept of the celestial sphere (see page 82) can help us explain what stars we can see and how they appear to move with respect to our particular

viewing location. The latitude of the observation point is the key: a skywatcher in, for example, Canada sees the stars with much the same perspective as someone in central Europe or in northern Japan; the same is true for observers in southern latitudes, such as Buenos Aires, Cape Town and Adelaide.

At the north pole, 90° N For an observer at the Earth's north pole, the celestial north pole, marked by the bright star Polaris in Ursa Minor, corresponds with the zenith—the point in the sky directly overhead.

From this position on Earth, the celestial equator is parallel with the horizon, and because the stars move along a path parallel to this horizon, only the stars in the northern half of the celestial sphere are ever visible.

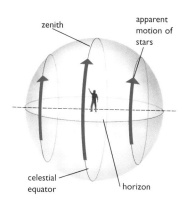

AT THE EQUATOR, 0°
All of the stars in the sky are visible at some time to an observer at the equator.

At the equator, 0° From a point anywhere on the equator, an observer can see all of the stars in the sky. At this latitude, the celestial equator rises up in an imaginary vertical line from the east point on

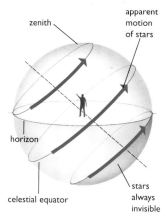

AT 40°S LATITUDE
Part of the sky is always invisible to observers in most parts of the world.

the horizon and runs through the zenith. The north and south celestial poles lie exactly on the horizon. The stars rise straight up in the east and later sink straight down below the horizon in the west.

At 40°S latitude For all locations lying between these latitudes, there is a part of the sky that always remains invisible—that which surrounds the celestial pole of the opposite hemisphere. On the other hand, the area of the sky close to the visible pole remains in view all the time. Stars here never set, but seem to circle around the celestial pole. These are called circumpolar stars, and the closer you are to the pole, the more circumpolar stars there are.

TIME AND THE SEASONS

While the tilted Earth spins around its own axis, it is also tracing out a path around the Sun. These two motions determine our system of timekeeping and the succession of the seasons. Our planet's yearly progress around the Sun also produces gradual changes in the positions of the stars.

SLOW MOVEMENT...
As the Earth moves around the Sun, stars appear to move. The illustration shows the constellation Orion, as seen looking south in winter from Edinburgh, Scotland.

...BRINGS GRADUAL CHANGE
Two weeks later at the same place and same time of night, Orion has moved slightly to the west. Every night, any given star will rise about four minutes earlier.

Solar and sidereal time Our day of 24 hours is how long it takes the Earth to complete one rotation around its axis relative to the Sun. This is *mean solar time*. Relative to the stars, the Earth rotates once in 23 hours and 56 minutes. The difference is due to the slowly changing view as the Earth moves around its orbit. Because (for our purposes) the stars are essentially fixed in space, this is the Earth's true period of rotation and is the basis of *sidereal time*.

The seasons Why we experience seasons has to do with the fact that the Earth is tilted relative to the plane of its orbit around the Sun. Thus, solar rays strike the ground at ever-changing angles during the course of a year. The varying amounts of solar energy produce the seasonal changes we experience.

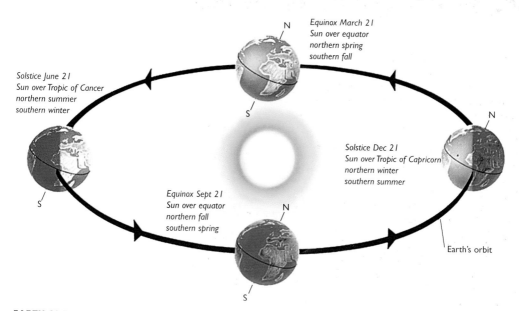

Solstice June 21
Sun over Tropic of Cancer
northern summer
southern winter

Equinox March 21
Sun over equator
northern spring
southern fall

Solstice Dec 21
Sun over Tropic of Capricorn
northern winter
southern summer

Equinox Sept 21
Sun over equator
northern fall
southern spring

Earth's orbit

EARTH AND THE SEASONS

Seasons are caused by the Earth's tilt on its axis of 23.5 degrees from the vertical. In June, the Southern Hemisphere is in winter because it is leaning away from the Sun, while the Northern Hemisphere is experiencing midsummer. Six months later, the Northern Hemisphere is in midwinter, while the Southern Hemisphere is in midsummer.

Understanding the Sky

FINDING YOUR DIRECTION

How do you find your way to a particular place in a sky that is so full of stars? The trick is to orientate yourself by finding either north or south. You can then go on to identify bright stars or constellations that will act as jumping-off points.

North or south? Observers in the Northern Hemisphere need to find north; in the Southern Hemisphere, find south. (See the diagrams on the opposite page.)

Landmarks A few stars and constellations are easy to find. Start off at one of these and move around the sky, a star or constellation at a time. Landmark constellations include: Ursa Major, which includes the Big Dipper; the W-shape of Cassiopeia; Orion; the Great Square of Pegasus; Leo with its sickle; Scorpius; and Crux, the Southern Cross.

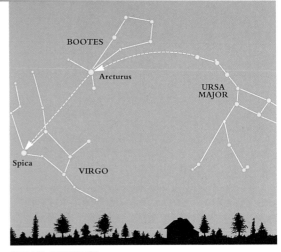

ARC TO ARCTURUS
The Big Dipper can lead you to other stars and constellations in the northern sky. Joining the three stars in the handle forms a curved line or arc. If you extend the line away from the Dipper you can "arc to Arcturus," the brightest star in Boötes, the Herdsman, and then "speed to Spica," in the same direction—the bright star in Virgo. You can devise your own star-hopping journeys. Start at a bright star and move around the sky, one star at a time, until you reach your destination.

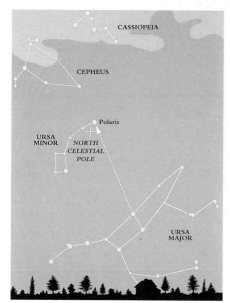

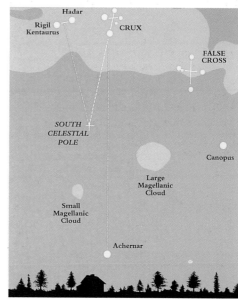

FINDING NORTH

North lies near Polaris. To find it, locate the Big Dipper (in Ursa Major), mentally draw a line joining the two stars at the end of the bowl, then extend it five times.

FINDING SOUTH

Southern Hemisphere observers can find the south celestial pole by extending the long arm of Crux, the Southern Cross, four and a half times to reach close to the pole.

STAR BRIGHTNESS AND COLOR

Watch the stars on a clear night and, as your eyes grow accustomed to the dark, you will notice how much stars vary in brightness. They do so for two reasons: some are closer to us than others, and some really are brighter than others. The concepts of absolute and apparent magnitude allow us to describe star brightness precisely. Stars also differ in color, giving us a clue to their nature.

Apparent magnitude This indicates how bright a star appears to the naked eye—the lower the magnitude, the brighter the star. For example, the brightest star in the sky, Sirius in Canis Major, shines at −1.5 magnitude, while the faintest stars visible to the eye in a dark sky are about magnitude 6. The scale is such that a 1st magnitude star is 100 times brighter than a 6th magnitude star. Most references in this book to a star's brightness concern its brightness in the sky as seen from Earth—apparent magnitude.

ILLUSION VERSUS REALITY
The stars of Ursa Major form the familiar Big Dipper (or Plough) purely by chance since they actually lie at different distances from the Earth.

Absolute magnitude The nearer a star is, the brighter it will appear, and since different stars lie at different distances, apparent magnitude does not measure the true brightness of a star. In order to describe intrinsic brightness, astronomers have defined *absolute* magnitude as the apparent magnitude a star would have if it were 10 parsecs or about

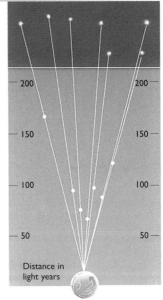

200
150
100
50

200
150
100
50

Distance in light years

33 light years from us. For example, the Sun's apparent magnitude is –26.8, but its absolute magnitude is +4.8. On the other hand, the absolute magnitude of Sirius is +1.4. So, Sirius is intrinsically 3.4 magnitudes more luminous than the Sun.

Star colors If you look at Antares in Scorpius, Aldebaran in Taurus, or Betelgeuse in Orion, you will see that all these stars are reddish. Vega in Lyra and Rigel in Orion are a delicate blue-white. A star's color is an indication of its temperature. Blue-white signifies that the star is much hotter than the Sun; red means it is cooler. Star colors are subtle to the eye. In photos the colors appear more obvious, but they are also influenced by the film's sensitivity.

Color and the eye The eye itself has a particular color sensitivity—we see yellow and green clearly, but infrared or ultraviolet (UV) light is invisible. The type of magnitudes we usually encounter are more properly called visual (V) magnitudes since they reflect the brightness of stars as seen by the human eye. Yet some stars are very dim in visible light but bright in, say, the infrared (which we can't see). Our visual magnitude system provides a poor reflection of the true energy output of such stars.

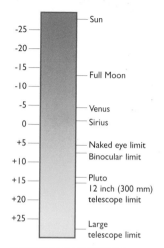

APPARENT MAGNITUDE SCALE
The apparent magnitude scale describes how objects *appear* in our sky; it does not describe the objects' true brightness.

STAR COLOR
A star's color is related to its surface temperature. The color is subtle, and most stars look white to the eye.

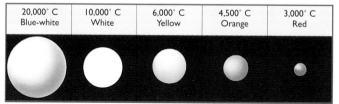

20,000° C Blue-white	10,000° C White	6,000° C Yellow	4,500° C Orange	3,000° C Red

91

OBSERVING THE PLANETS

Planets can be divided into two groups. The interior planets—Venus and Mercury—circle the Sun inside the Earth's orbit. Exterior planets are those from Mars outward; they stay outside of Earth's orbit. The difference matters because it determines how and where you need to look for a planet.

VENUS
This view through a telescope shows the crescent disk of Venus. The planet's clouds are almost as reflective as snow, which is why Venus looks so bright to the naked eye.

CRESCENT NEPTUNE
Neptune, as photographed by the Voyager 2 spacecraft, illustrates how a planet that is closer to the Sun than the observer can show a crescent phase.

Interior planets An interior planet always stays near the Sun in the sky; you will never see one at midnight, for example. As Mercury or Venus orbits the Sun, you may see it first in the evening sky, moving day by day out from the Sun. After a few weeks, it reaches a point of maximum separation from the Sun called greatest eastern elongation, the time of best visibility. The planet passes between the Sun and the Earth at a point called inferior conjunction, when it briefly disappears in the solar glare. Greatest western elongation follows in the pre-dawn, bringing good visibility. The planet completes its orbit on the far side of the Sun, and disappears again. This ends one apparition, or viewing season, and begins the next.

INTERIOR-EXTERIOR
Interior planets reach best visibility at the points of greatest elongation; exterior planets are best seen at opposition.

Exterior planets We first notice an exterior planet rising just ahead of the Sun, low in the pre-dawn sky. As weeks pass, it rises earlier each night. Many weeks before opposition, the planet's motion against the background stars stops, and then it appears to move backward (westward). This retrograde motion is caused by the Earth's greater orbital speed. At opposition the planet reaches a point where it rises as the sun sets. This is the time of best visibility. After several weeks, it stops moving retrograde and recommences eastward motion. It continues to rise earlier each day, providing fine evening viewing. Eventually the planet disappears into the Sun's glare and reaches conjunction.

Finding planets Planets orbit close to the ecliptic, the path of the Sun in the sky. This makes them easier to find—the ecliptic is usually marked on sky charts. But unlike constellations, whose seasons of visibility can be predicted easily, planets are on the move. Astronomy magazines provide regular monthly guides on planetary positions, and annual handbooks and almanacs do likewise on a yearly basis.

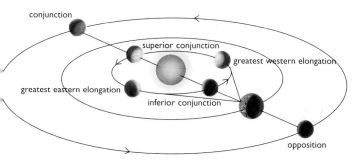

conjunction

superior conjunction

greatest western elongation

greatest eastern elongation

inferior conjunction

opposition

COLUMBA · CAELUM · HOROLOGIUM · CETUS · Canopus · DORADO · PICTOR · RETICULUM · CARINA · Turantula Nebula · MENSA · PHOENIX · SCULPTOR · Achernar · VOLANS · LMC · HYDRUS · SMC · CHAMAELEON · 47 Tuc · TUCANA · GRUS · MUSCA · OCTANS · APUS · INDUS · PISCIS AUST. · Fomalhaut · Acrux · Jewel Box · TRIANGULUM AUST. · Beta Cen · PAVO · CIRCINUS · Alpha Cen · Omega Centauri

SOUTH

URSA MINOR · +90° Polaris · URSA MAJOR

SKY
AND
CONSTELLATION CHARTS

S tar charts may look chaotic or intimidating at first glance, but they contain a lot of order and are surprisingly easy to use. After all, they are simply maps of the night sky. As you gain experience in their use, you will find that they become as familiar to you as any roadmap. This chapter contains two types of star charts: the sky charts give you the big picture, showing the night sky's constellations season by season for both hemispheres; the constellation charts contain much more detail, zooming in on specific star patterns and their brightest deep-sky objects of interest. Also included are all the information and instructions you need to make the most of the charts.

USING THE SKY CHARTS

The sky charts on pages 100 to 115 show the sky's brightest stars, which are joined together as constellations. There are 16 sky charts in all—eight for Northern Hemisphere observers and eight for those in the Southern Hemisphere. Each is in two parts on facing pages—the left-hand chart looking north, the right-hand chart looking south.

HOW STARS ARE NAMED

The names we use for stars come mainly from the early Greek and Arab astronomers. But several other naming systems are also used.

Bayer letters This system applies the Greek alphabet to stars in a constellation, so that the brightest star is generally called alpha (α), the next brightest beta (β), and so on. Betelgeuse in Orion is therefore also called Alpha (α) Orionis, Orionis being the genitive form of the constellation name.

Flamsteed numbers Stars are numbered west to east across a constellation, so Betelgeuse is also 58 Orionis.

M objects These are star clusters, nebulae and galaxies in the Messier list, compiled by the eighteenth-century French comet-hunter Charles Messier.

NGC objects These are galaxies, star clusters and nebulae listed in the New General Catalogue of J. E. L. Dreyer, published in the late nineteenth century. The NGC, along with its two index catalogs (IC), lists over 13,000 objects.

Other systems The Smithsonian Astrophysical Observatory Catalog includes 258,997 stars. Another bigger and more recent list is the Hubble Space Telescope (HST) Guide Star Catalog which includes some 19 million stars.

About the charts Each chart shows the whole horizon between east and west, and runs from the horizon to straight overhead. Different-size dots indicate the stars and their apparent brightnesses. The charts show the evening sky in each season for observers in middle northern or southern latitudes. People at different latitudes will see the sky slightly shifted. The view will also vary a little according to the time of night and the date within each season.

Step 1 Identify north and south (see pages 88 and 89).

Step 2 Turn to the appropriate sky chart for your hemisphere and the current season (see pages 100 to 115). Pick either the north- or south-looking chart and read the accompanying description.

Step 3 Choose two or three of the brightest stars featured (the biggest dots on the chart) and try to find them in the sky. Remember to mentally adjust the scale of the chart to the real sky.

Step 4 Once you have identified one of these stars, attempt to trace out the constellation of which it is a part, then trace out the constellations nearby.

Step 5 For more information on a particular constellation, look up the appropriate constellation chart (see pages 116 to 285).

RED FLASHLIGHT
Use a a red-filtered flashlight to keep your eyes dark adapted during observing sessions, as any bright white light will close the eyes' pupils again.

25 BRIGHTEST STARS		
Common name	**Constellation name**	**App. mag.**
Sirius (d)	α Canis Majoris	-1.46
Canopus	α Carinae	-0.72
Alpha Centauri (d)	α Centauri	-0.01
Arcturus	α Boötis	-0.04
Vega	α Lyrae	0.03
Capella	α Aurigae	0.08
Rigel	β Orionis	0.12
Procyon (d)	α Canis Minoris	0.8
Achernar	α Eridani	0.46
Hadar (v)	β Centauri	0.66
Betelgeuse (v)	α Orionis	0.70
Altair	α Aquilae	0.77
Aldebaran	α Tauri	0.85
Acrux (v)	α Crucis	0.87
Antares (v)	α Scorpii	0.92
Spica (v)	α Virginis	1.00
Pollux	β Geminorum	1.14
Fomalhaut	α Piscis Austrini	1.16
Deneb	α Cygni	1.25
Beta Crucis (v)	β Crucis	1.28
Regulus	α Leonis	1.35
Adhara	ε Canis Majoris	1.50
Castor (d)	α Geminorum	1.59
Shaula (v)	λ Scorpi	1.62
Bellatrix	χ Orionis	1.64
(d) = double star (v) = variable star		

USING THE CONSTELLATION CHARTS

A multitude of amazing objects appear in the sky, and the constellation charts on pages 116 to 285 and their accompanying descriptions allow you to identify particular ones that may be of interest to you. Simply choose a constellation and start your search for individual stars, as well as clusters, galaxies and nebulae.

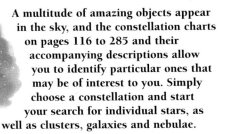

PLEIADES CLUSTER
The Pleiades Cluster forms the bull's shoulder in Taurus.

About the charts The charts are presented in alphabetical order, and with north at the top and east to the *left* (not the right, as on maps of the Earth). Within the constellation, stars down to magnitude 6.5 are shown; those which fall outside the constellation are stars down to magnitude 5.5. All the stars marked are therefore visible to the naked eye under dark skies, but in towns or cities, binoculars may be needed to pick out the fainter ones. (As a general rule-of-thumb, the appropriate magnitude limits are: cities—2 or 3; suburbs—4; distant suburbs—4.5 or 5; rural—5 to 6.5.) Many stars are named with letters of the Greek alphabet. These are shown clearly on the charts. In addition to the naked-eye stars, the charts mark the locations of other important objects such as star clusters, nebulae and galaxies; most of these require optical aids to be seen. Any deep-sky objects of particular interest are included, down to around magnitude 11.

Viewing Many objects of interest are within the range of binoculars or telescopes with apertures of 2.4 inches (60 mm). Of course, most objects will reveal more detail when viewed with a larger telescope.

Features Each of the constellation charts features the following:
- a pronunciation guide and the constellation's common name
- the genitive form of the constellation name, used when correctly naming objects within the constellation—alpha in Libra is alpha Librae, for example
- the three-letter abbreviation of the constellation's proper Latin name

• best viewing time, which is the approximate date the constellation is on meridian (that is, highest in the sky) at 10 p.m.—standard time, not daylight saving time

• a visibility grading, based on a scale from 1 to 4, representing the ease with which the constellation can be seen

• a hand symbol, which indicates the number of outstretched hand spans (each about 20 degrees across) that will cover the constellation east to west (left to right on the map)

• where applicable, a symbol indicating that one of the 25 brightest stars is featured in the constellation (see table, page 97)

Measurements The descriptions that accompany the charts give distances and sizes of objects in degrees, minutes and seconds. Ninety degrees is the distance from

CHART SYMBOLS							
Magnitudes	-1	0	1	2	3	4	5 6 and under
Double stars		Variable stars			Open clusters		
Globular clusters			Planetary nebulae				
Diffuse nebulae			Galaxies			Quasar	

the horizon to the zenith point directly overhead. The Moon and Sun present disks of about ½ degree, or 30 arc minutes, in size. The finest detail your eye can resolve without an optical aid is about 1 arc minute, or 60 arc seconds. You can use your hand held at arm's length to get a rough idea of distances (see the illustrations below).

TEXT SYMBOLS

 Object is first easily viewed with naked eye

 Object is first easily viewed with binoculars

 Object is first easily viewed with a telescope

 Visibilty grading

 Apparent size (hand spans)

 Contains one of the 25 brightest stars

JUDGING DISTANCES

To gain a rough idea of apparent distances, hold your hand at arm's length. With fingers spread, your hand covers approximately 20 degrees of sky; a fist covers about 10 degrees; and a thumb, 2 degrees.

NORTHERN HEMISPHERE—SPRING

Looking north The landmark of the northern sky is Ursa Major, the Great Bear. It lies tonight above the pole star, Polaris, which is in Ursa Minor, the Little Bear. Ursa Major's brightest stars form the Big Dipper. The two stars that form the left-hand side of the Dipper's bowl point down to Polaris. Follow the curving handle of the Dipper as it "arcs to Arcturus" in Boötes, the Herdsman, who sits high in the southeast.Standing in the northwest are Gemini, the Twins, with Castor and Pollux. In the northeast, bright Vega heralds the rise of Lyra, the Lyre, with Hercules above.

WEST   NORTH   EAST

Looking south As Procyon in Canis Minor, the Little Dog, edges westward, three bright stars light the southern sky—Arcturus in Boötes, the Herdsman (in the southeast); Spica in Virgo, the Maiden (south); and Regulus in Leo, the Lion (southwest). Below Leo and Virgo runs the night sky's longest constellation—Hydra, the Water Snake—but it is hard to see except on dark and moonless nights. Between Virgo and the tail of Leo lies the Realm of the Galaxies, the nearest cluster of galaxies to our own. This group of spiral and elliptical galaxies lies about 65 million light years away.

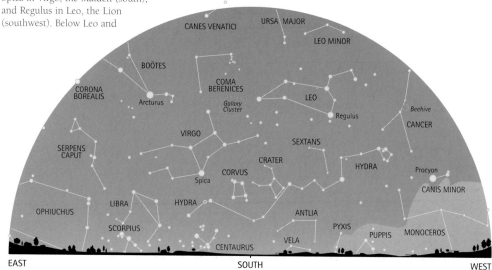

NORTHERN HEMISPHERE—SUMMER

Looking north The Big Dipper (in Ursa Major, the Great Bear) lies to the left of the pole star, Polaris, opposite the W-shape of Cassiopeia, the Queen, rising on the right. Above Ursa Minor, the Little Bear, look for the dim form of Draco, the Dragon, which wraps around the pole. If the night is moonless and dark, you can see the Milky Way span the sky from northeast to southwest, passing from Perseus, the Hero, through Cassiopeia and Cepheus, the King, to Cygnus, the Swan. Cygnus is also known as the Northern Cross. It appears as a bird flying south along the Milky Way.

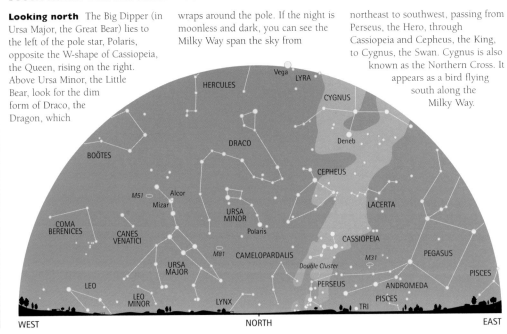

Looking south The richest part of the Milky Way parades across the south on summer nights, featuring the constellations of Scorpius, the Scorpion, and Sagittarius, the Archer. (Sagittarius resembles the profile of a teapot, with Milky Way "steam" rising from its spout.) Here lies the mysterious center of the Milky Way Galaxy, hidden behind 30,000 light years of dusty gas. Farther north, the Milky Way divides alongside Aquila, the Eagle, and Cygnus, the Swan, because a cloud of interstellar dust near the Sun is obscuring the distant stars. You will see it best on a moonless night away from city lights.

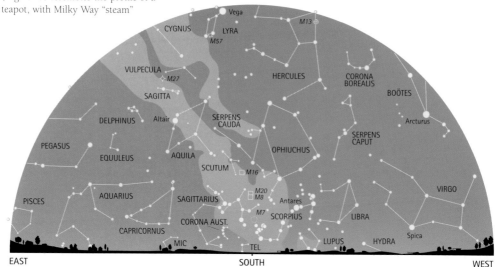

Sky and Constellation Charts

NORTHERN HEMISPHERE—FALL

Looking north The Milky Way spans the sky for those viewing away from streetlights on a moonless night. To the north, the bright stars of the Big Dipper in Ursa Major, the Great Bear, graze the horizon, while Cassiopeia, the Queen, makes a bent M-shape above Polaris, the pole star. As Auriga, the Charioteer, rises in the northeast, the Summer Triangle of the stars Vega, Deneb and Altair is descending in the west. High overhead lies the great Andromeda Galaxy (M 31)—the closest major galaxy to us. It is the most distant object visible to the naked eye, at 2.9 million light years away.

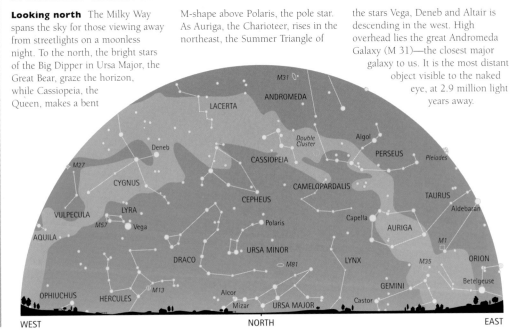

WEST   NORTH   EAST

Looking south High in the south, the Great Square of Pegasus is a notable landmark. The Flying Horse is upside down, so the line of stars to the lower right of the Square delineates his neck. This line reaches west toward the Milky Way and the bright star Altair in Aquila, the Eagle. Look below the Great Square to spot various "watery" constellations: Pisces, the Fish; Cetus, the Whale; Aquarius, the Water Bearer; and Piscis Austrinus, the Southern Fish. In the eastern sky, Taurus, the Bull, is making his appearance, followed by winter's Orion, the Hunter, rising out of the eastern horizon.

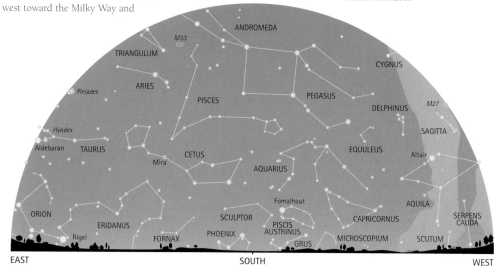

NORTHERN HEMISPHERE—WINTER

Looking north The Big Dipper in Ursa Major, the Great Bear, stands on its handle, while the two top bowl stars point left to Polaris in Ursa Minor, the Little Bear. As the Dipper rises, Cassiopeia, the Queen, sinks, along with her mythological companions—Cepheus (her husband), Andromeda (her daughter), Perseus (Andromeda's hero and savior) and Pegasus (Perseus' horse). Appropriately, Cetus, the Whale, who was about to devour Andromeda when Perseus intervened, has already set. In the east, Leo, the Lion, is rising, marked by the 1st magnitude star Regulus.

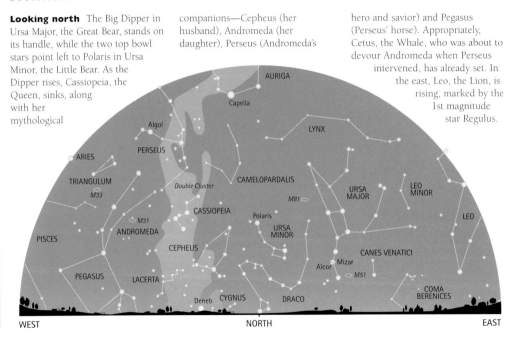

WEST   NORTH   EAST

Looking south Orion, the Hunter, strides across the horizon, driving back Taurus, the Bull, with the lovely Pleiades star cluster on its back. Taurus' face is formed from a looser star cluster, the Hyades. To the lower left of Orion, Canis Major, the Great Dog, follows the Hunter, with Sirius (the brightest star in the sky) marking his eye. Left of Orion, the star Procyon distinguishes Canis Minor, the Little Dog, while Castor and Pollux stand at the head of Gemini, the Twins. Alongside Gemini is Auriga, the Charioteer, wheeling overhead along the Milky Way.

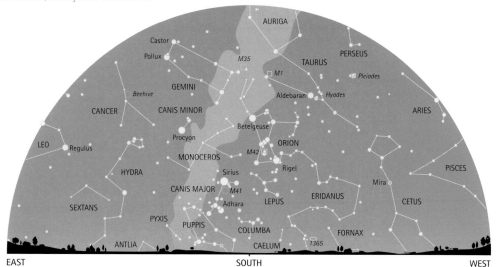

EAST SOUTH WEST

SOUTHERN HEMISPHERE—SPRING

Looking north The Great Square of Pegasus, the Winged Horse, forms a landmark as he gallops west across the northern horizon. The two lower stars in the Square point west toward bright Altair in Aquila, the Eagle. The two stars on the left side of the Square

point up to Fomalhaut in Piscis Austrinus, the Southern Fish, high overhead. Between Pegasus and

Fomalhaut lie many dim "watery" constellations: Pisces, the Fish; Cetus, the Whale; and Aquarius, the Water Bearer. Rising in the east is Aldebaran—the ruddy eye of Taurus, the Bull—with Orion, the Hunter, on the horizon.

Looking south Two bright stars make a triangular shape with the south celestial pole: Achernar in Eridanus, the River, and Canopus in Carina, the Keel. The pole lies about where the lower right star of the triangle would be. Between

Canopus and Achernar lies the Large Magellanic Cloud (LMC), a satellite galaxy of the Milky Way, with the Small Magellanic Cloud (SMC) slightly above it. Several celestial

birds flock here: Phoenix, above Achernar; Grus, the Crane, to Phoenix's right; and Tucana, the Toucan, and Pavo, the Peacock, below them both. The Milky Way lines the western horizon.

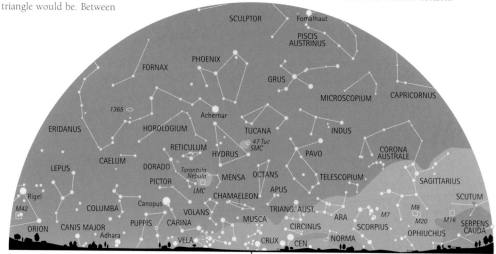

EAST   SOUTH   WEST

SOUTHERN HEMISPHERE—SUMMER

Looking north The tall figure of Orion, the Hunter, with his three-starred Belt, stands high in the north. Extending the Belt down to the left points toward Taurus, the Bull, with ruddy Aldebaran and the Hyades and Pleiades. Extending the Belt

upward to the right points to Sirius, the sky's brightest star, in Canis Major, the Great Dog. From Sirius,

a line down to the northeast passes Procyon in Canis Minor, the Little Dog, to Regulus in Leo, the Lion. Below Orion, yellow Capella arcs low with Auriga, the Charioteer. To its right stand the bright Twins of Gemini— Castor and Pollux.

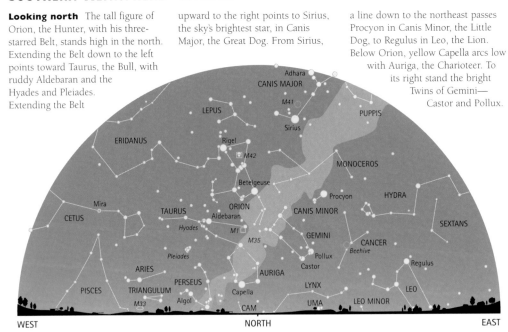

WEST　　　　NORTH　　　　EAST

Looking south High in the southern sky the bright star Canopus is the rudder in Carina, the Keel. The rest of the ship is made up by Vela, the Sails, and Puppis, the Stern, both of which lie in the Milky Way. Rising in the southeast below Vela is tiny Crux, the Southern Cross, with Centaurus, the Centaur, wrapped around it. During this season, the southwestern sky has a single bright star, Achernar, a beacon at the end of Eridanus, the River.

Look between Achernar and Canopus for a misty patch—the Large Magellanic Cloud (LMC). It forms a triangle with the Small Magellanic Cloud (SMC) and Achernar.

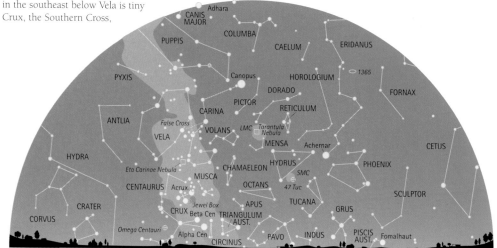

EAST   SOUTH   WEST

SOUTHERN HEMISPHERE—FALL

Looking north Four bright stars make it easy to find constellations this season. In the northwest, look for Procyon in Canis Minor, the Little Dog. In the north, spot Regulus in Leo, the Lion. Then in the northeast, warm-tinted Arcturus stands in Boötes, the Herdsman, with white Spica in Virgo, the Maiden, above. Between Leo and Virgo lies the Virgo cluster—the nearest cluster of galaxies to our own, at about 65 million light years away. High overhead, the long figure of Hydra, the Water Snake, has his head close to Procyon while his body weaves past Crater, the Cup, and Corvus, the Crow.

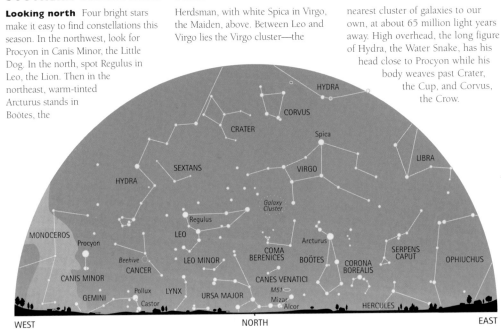

Looking south The glories of the Milky Way arch across the south during spring, from Sirius in Canis Major, the Great Dog, setting in the west, to ruddy Antares in Scorpius, the Scorpion, rising in the southeast. Above Canopus in the southwest lies Carina, the Keel; Vela, the Sails; and Puppis, the Stern—all part of Jason's mythical ship *Argo*. To the left of these stands tiny Crux, the Southern Cross, which Centaurus, the Centaur, so nimbly hops over, his forefeet marked by Alpha and Beta Centauri. If possible, get away from city lights and explore the Milky Way with a pair of binoculars.

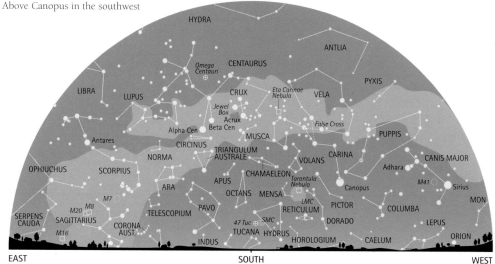

SOUTHERN HEMISPHERE—WINTER

Looking north Three bright stars line the northern horizon. In the northwest, Arcturus sets with Boötes, the Herdsman. Due north stands Vega, hallmark of Lyra, the Lyre; higher in the northeast, in the Milky Way, is Altair in Aquila, the Eagle.

(Beneath them, you may spot Deneb in Cygnus very low in the sky.) Right of Boötes lie Corona Borealis, the

Northern Crown, and Hercules. If the night is dark and moonless, look for the great rift in the Milky Way near Aquila, caused by a huge cloud of dust. The Milky Way widens overhead because there lies the populous heart of our galaxy.

WEST   NORTH   EAST

Looking south The Milky Way is setting in the southwest and taking with it some of the brightest stars. Crux, the Southern Cross, is easy to identify, while Centaurus straddles it. Alpha and Beta Centauri mark the Centaur's forelegs. Above the Centaur creeps Lupus, the Wolf, and a star-rich run of Milky Way that reaches to Scorpius, the Scorpion, high overhead. Here, and in adjoining Sagittarius, the Archer, lie many of the Milky Way's greatest sights. On the next moonless night, find an observing site away from city lights and explore this region with binoculars or a telescope.

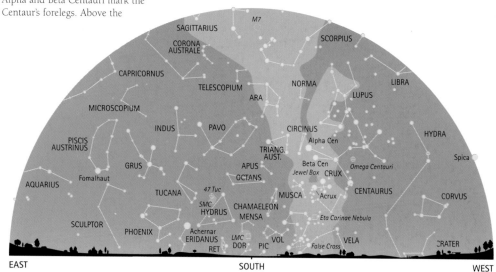

ANDROMEDA (an-DROH-me-duh)
The Chained Princess

Although Andromeda is renowned for the great galaxy that resides within the constellation, its stars are not very bright. It is easy to find, however, located south of Cassiopeia's W, and just off one corner of the Great Square of Pegasus. In fact, Alpheratz, the star that sits at the northeastern corner of the square of Pegasus, belongs to Andromeda. In mythology, Andromeda was the daughter of Cassiopeia and Cepheus, rulers of the ancient country of Æthiopia.

The Andromeda Galaxy (M 31) The closest major galaxy to us, the Andromeda Galaxy was first thought to be a nebula, and was listed in comet hunter Charles Messier's eighteenth-century catalogue of nebulae. A spiral galaxy much like our own Milky Way, it is a maelstrom comprising 200 billion stars and clouds of dust and gas. It is bright enough to be seen with binoculars from city sites and with the naked eye beneath a dark sky, being one of the most distant objects visible to the unaided eye. In the field of larger binoculars, or using a small telescope, you can see its two neighboring elliptical galaxies. M 32 is small and compact; M 110 is larger and more diffuse, and is therefore harder to see.

Gamma (γ) Andromedae This is a beautiful double star. The brighter member of the pair is a golden yellow, and its companion is greenish blue.

R Andromedae This Mira star has a range of 9 magnitudes.

NGC 752 This open cluster lies about 5 degrees south of Gamma (γ) Andromedae and is easy to find because of its relatively bright stars. Because it is spread out over such a large area, it is actually easier to see through binoculars than through a telescope. If using a telescope, use it at its lowest power.

NGC 7662 A fairly bright planetary nebula, this blue-green object looks almost starlike through the smallest telescopes. But through a 6 inch (150 mm) telescope at moderate power, it becomes a graceful, glowing spot of gas about 30 arc seconds across.

NGC 891 This galaxy is a challenge even for 6 inch (150 mm) telescopes. However, with a dark sky, you will see one of the best examples of a spiral galaxy, viewed edge-on.

ANDROMEDAE (AND)
On meridian
10 p.m. Nov 1

KEY

x 2

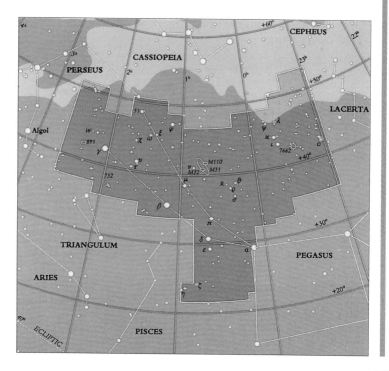

ANTLIA (ANT-lee-uh)
The Air Pump

Antlia Pneumatica, the Air Pump, named after seventeenth-century physicist Robert Boyle's invention, is a southern constellation. It was given this somewhat unpoetic name by Nicolas-Louis de Lacaille during the time he spent working at an observatory at the Cape of Good Hope, from 1750 to 1754. As a result of his observations of some 10,000 southern stars, de Lacaille divided the far southern sky into 14 new constellations, of which Antlia is one.

Put three grains of sand inside a vast cathedral, and the cathedral will be more closely packed with sand than space is with stars.

SIR JAMES JEANS (1877–1946),
English astronomer

NGC 2997 This is a large, faint spiral galaxy, with a stellar nucleus. It is quite difficult to observe with a small telescope.

SEEING ANTLIA

Antlia is a small, faint constellation just off the bright southern Milky Way, not far from Vela and Puppis. Its alpha (α) star is just barely the constellation's brightest star and has been given no proper name. It is red in color and possibly varies slightly in magnitude.

ANTLIAE (ANT)
On meridian
10 p.m. March 20

KEY

x 1

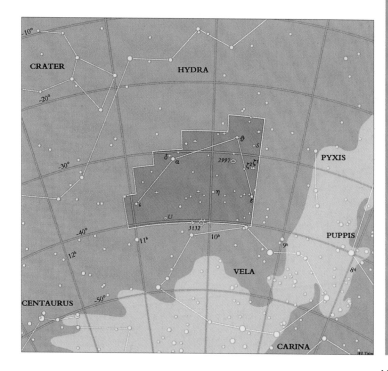

APUS (ay-pus)
The Bird of Paradise

This faint constellation is directly below Triangulum Australe, the Southern Triangle. Being close to the southern pole, it cannot be seen from most northern latitudes. *Apus* is an ancient Greek word that means "footless" and derived from *Apus indica*, the name given to India's Bird of Paradise. This magnificent bird was offered as a gift to Europeans, but not before its unsightly legs were cut off.

S Apodis This is a "backward" nova. Usually the star shines slightly brighter than magnitude 10, bright enough to see through a small telescope, but at irregular intervals it erupts, possibly sending dark, sootlike material into its atmosphere. It then fades dramatically by some 100 times, to about magnitude 15. After staying faint for several weeks, it slowly returns to its original brightness.

Theta (θ) Apodis This variable star ranges from magnitude 6.4 to below 8 in a semi-regular cycle over 100 days.

NGC 6101 A faint globular cluster, large and slightly irregular, NGC 6101 can be seen as a small, misty spot through a small telescope.

> Stars are like animals in the wild. We may see the young but never the actual birth, which is a veiled and secret event.
>
> HEINZ R. PAGELS (1939–88),
> American physicist

APODIS (APS)
On meridian
10 p.m. June 20

KEY

x 1.5

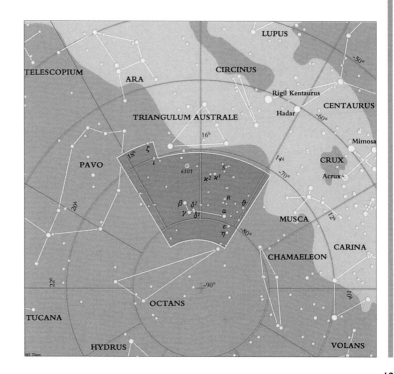

AQUARIUS (ah-KWAIR-ee-us)
The Water Bearer

The Water Bearer dates as far back as Babylonian times and is appropriately placed in the sky not far from a dolphin, a river, a sea serpent and a fish. Of its many mythological associations, it was at times identified with Zeus, the chief Greek god, pouring the waters of life down from the heavens.

M 2 This fine globular cluster appears as a fuzzy spot of light through binoculars and small telescopes. It is possible, however, to see the cluster's mottled appearance through a 4 inch (100 mm) telescope, and to resolve it into stars through a 6 inch (150 mm) telescope.

The Saturn Nebula (NGC 7009) This small planetary nebula was named by Lord Rosse who, with his large reflecting telescope, first saw the protruding rays that made it look like a dim version of Saturn with its rings. It is visible through a telescope as a greenish point of light.

The Helix Nebula (NGC 7293) The largest and closest (at 450 light years away) of the planetary nebulae, this cloud takes up half the angular diameter of the Moon in the sky. Because its brightness is spread over a large area, it appears best with a low-power, wide-field telescope or binoculars under a dark sky.

Delta (δ) Aquarids This strong meteor shower peaks every year on July 28, with meteors seeming to emerge from a point in Aquarius.

AQUARII (AQR)
On meridian
10 p.m. Sept 20

KEY

x 2.5

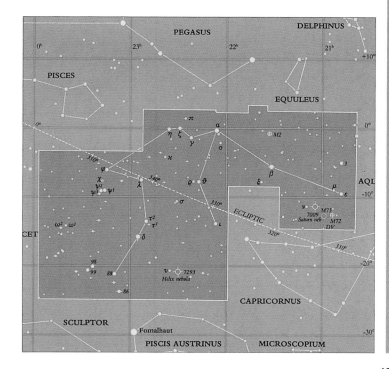

AQUILA (uh-KWI-luh)
The Eagle

Identified in ancient times as an eagle by astronomers of the Euphrates Basin, the constellation of Aquila takes its name from the bird that belonged to the Greek god Zeus. Aquila's main accomplishment was to bring the handsome mortal youth Ganymede to the sky to serve as his master's cup bearer.

👁 **Eta (η) Aquilae** This supergiant star is a bright Cepheid variable that changes over a magnitude in brightness (3.5 to 4.4) in a period of little more than a week. At its brightest it rivals Delta (δ) Aquilae, and it fades to about the magnitude of Iota (ι) Aquilae.

R Aquilae This Mira star varies in magnitude from 6 to 11.5 over a period of 284 days.

NGC 6709 This pretty open cluster consists of a group of closely knit stars against an already rich background of stars.

AQUILAE (AQL)
On meridian
10 p.m. Aug 10

KEY

x 1.5

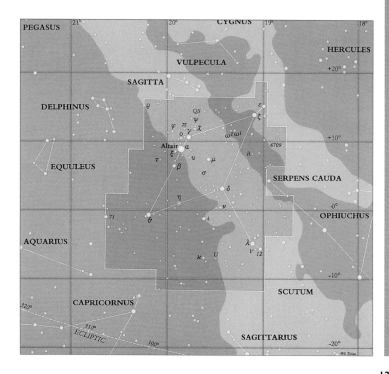

ARA (AR-uh)
The Altar

Located south of Scorpius, Ara's original Latin name was Ara Centauri—the altar of the centaur Chiron. Half man and half horse, Chiron was thought to be the wisest creature on Earth. Ara has also been referred to variously as the altar of Dionysus; the altar built by Noah after the flood; the altar built by Moses; and even the one from Solomon's Temple.

That the sky is brighter than the earth means little unless the earth itelf is appreciated and enjoyed. Its beauty loved gives the right to aspire to the radiance of the sunrise and the stars.

MY RELIGION,
HELEN KELLER (1880–1968),
American writer

U Arae This Mira-type variable is bright enough to be seen through a small telescope when it is at its maximum of magnitude 8. However, it then drops a full five magnitudes before rising again over a period of more than seven months.

NGC 6397 Possibly the closest globular star cluster to us, this bright cluster is placed between Beta (β) Arae and Theta (ϑ) Arae. The cluster is relatively loose, so an observer with powerful binoculars should be able to detect it without difficulty and perhaps even resolve its faint stars. It may be as large as 50 light years across.

ARAE (ARA)
On meridian
10 p.m. July 10

KEY

x 1

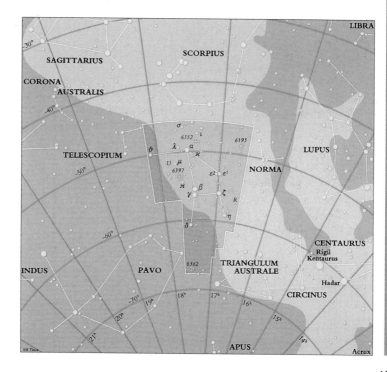

ARIES (AIR-eez)
The Ram

In a Greek legend, the king of Thessaly had two children, Phrixus and Helle, who were abused by their stepmother. The god Hermes sent a ram with a golden fleece to carry them to safety on its back. Helle fell off the ram as it was flying across the strait that divides Europe from Asia, a body of water the Greeks called the Hellespont, the sea of Helle (now the Dardanelles). Phrixus was carried to safety on the shores of the Black Sea, where he sacrificed the ram and its fleece was placed in the care of a sleepless dragon.

FIRST OF THE ZODIAC

Aries is the zodiac's first constellation, since the Sun at one time was entering Aries on the day of the vernal equinox—the moment when it crosses from the southern to the northern half of the celestial sphere. However, because of the Earth's precession, the Sun is now in Pisces at the vernal equinox. Aries is well known and is not difficult to find, but it has few objects of interest.

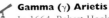

Gamma (γ) Arietis

In 1664, Robert Hooke was following the motion of a comet when he chanced upon this beautiful double star. One of the earliest doubles to be found with a telescope, this star has a separation of 8 arc seconds, and is easy to find and observe.

ARIETIS (ARI)
On meridian
10 p.m. Nov 20

KEY

x 1

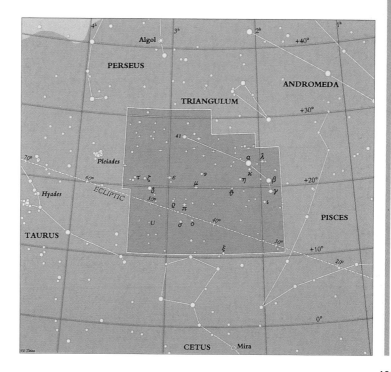

AURIGA (oh-RYE-gah)
The Charioteer

This multi-sided figure is easy to find in the sky, largely because of bright Capella, the she-goat star, and her retinue of three little kids. Ancient legends portray Auriga as a charioteer carrying a goat on his shoulder and two or three kids on his arm. The charioteer is also seen as Erechtheus, the son of Hephaestus (the Roman god Vulcan), who invented a chariot to move his crippled body about. Capella has been seen as the she-goat star since Roman times. Almost 50 light years away, Capella is a golden-yellow giant, larger than our Sun.

Epsilon (ε) Aurigae
An extraordinary variable system, this supergiant star fades when its companion passes in front of it once every 27 years. During an eclipse, its brightness drops by two-thirds of a magnitude. The deepest phase of the eclipse lasts a full year, which may indicate that the companion is surrounded by an enormous disk of gas and dust.

M 36 This bright open star cluster is some 5 degrees southwest of Theta (ϑ) Aurigae, and contains about 60 stars of 8th magnitude and fainter.

M 37 This is an exceptional open star cluster, almost the size of the Moon, and one of the finest in the northern sky. Binoculars will show this cluster as a misty spot. A small telescope will reveal its large number of stars.

M38 This small cluster of stars resembles the Greek letter π (pi) when seen in a small telescope.

AURIGAE (AUR)
On meridian
10 p.m. Dec 20

KEY

x 2

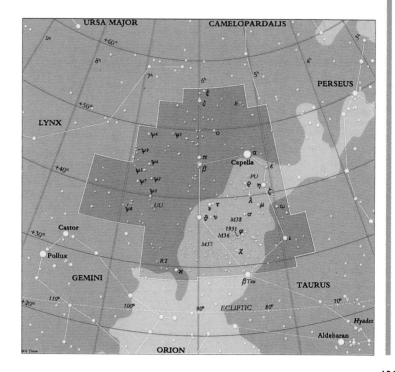

131

BOÖTES (boh-OH-teez)
The Herdsman

Boötes, whose name is derived from the Greek word for herdsman, was the son of Demeter. He is said to have been rewarded with a place in the sky for inventing the plow. Another legend tells of Boötes (also known as Arcas and Arcturus), son of Zeus and Callisto. Callisto, changed into a bear by Hera, the jealous wife of Zeus, was almost killed by her son when he was out hunting. Zeus rescued her, taking her into the sky where she became Ursa Major, the Great Bear. The name Arcturus (the constellation's brightest star) comes from the Greek meaning "guardian of the bear."

👁 **Arcturus** Alpha (α) Boötis, this yellow-orange star is 37 light years away from us, making it one of the closest of the bright stars. Arcturus's actual position in the sky has changed by over twice the Moon's apparent diameter in the last 2,000 years; astronomers say that Arcturus has a large proper motion.

FINDING BOÖTES
If you joint the three stars in the handle of the Big Dipper and "arc to Arcturus," you will find this constellation.

BOOTIS (BOO)
On meridian
10 p.m. June 1

KEY

x 1.5

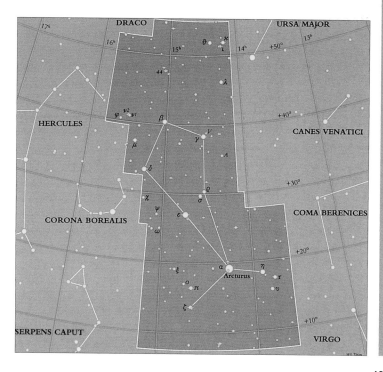

CAELUM (SEE-lum)
The Chisel

One of the least conspicuous of all the constellations, Caelum is one of the many regions in the Southern Hemisphere skies that was named by eighteenth-century astronomer Nicolas-Louis de Lacaille. It comprises a largely empty region of the heavens between the constellations of Columba, the Dove, and Eridanus, the River.

R Caeli A bright Mira-type variable, this star changes from magnitude 6.7 to 13.7 over a period of about 13 months.

*I stood upon that silent hill
And stared into the sky until
My eyes were blind with stars
and still
I stared into the sky.*

THE SONG OF HONOUR,
RALPH HODGSON (1871–1962),
British poet

CAELI (CAE)
On meridian
10 p.m. Jan 1

KEY

④

x 0.5

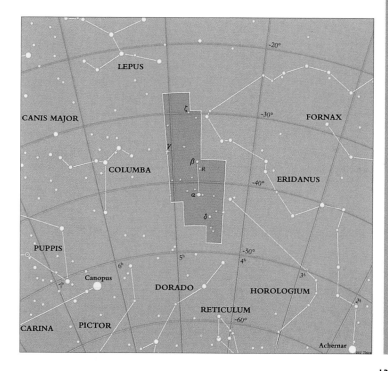

CAMELOPARDALIS (ka-mel-o-PAR-da-lis)
The Giraffe

What is a giraffe doing next to two bears and a dragon in the sky near Polaris? Camelopardalis was dreamed up by Bartsch in 1624, who claimed that it represented the camel that brought Rebecca to Isaac. ("Camel-leopard" was the name the Greeks gave to the giraffe, as they thought it had the head of a camel and a leopard's spots.) The constellation lies in the large space between Auriga and the bears.

Z Camelopardalis This cataclysmic variable star erupts every two or three weeks from its minimum of magnitude 13 to a maximum of 9.6, which is still quite faint. Its resemblance to other such variables ceases when, while fading, it stops changing and hovers at an intermediate magnitude. This "standstill" might last for months before the decline resumes. In the late 1970s, Z Cam stayed around magnitude 11.7 for several years.

VZ Camelopardalis This star varies irregularly over the small range between magnitudes 4.8 and 5.2. Located close to Polaris, VZ Cam is visible every night of the year from most northern latitudes.

*I open the scuttle at night and
   see the far-sprinkled systems,
And all I see multiplied as high
   as I can cipher edge but the
   rim of the farther systems.
Wider and wider they spread,
   expanding, always
   expanding,
Outward and outward and
   forever outward.*

SONG OF MYSELF,
WALT WHITMAN (1819–92),
American poet

CAMELOPARDALIS (CAM)
On meridian
10 p.m. Jan 10

KEY

x 2

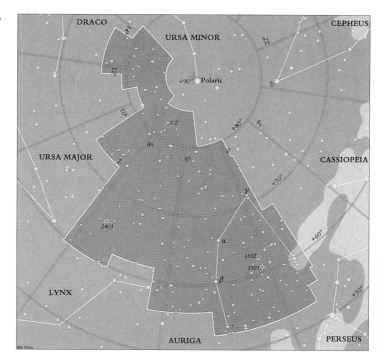

CANCER (CAN-ser)
The Crab

In Greek myth, Cancer was sent to distract Hercules when he was fighting with the monster Hydra. The crab was crushed by Hercules's foot, but as a reward for its efforts Hera placed it among the stars. Cancer lies between Gemini and Leo—two of the sky's showpieces. It has no star brighter than 4th magnitude and its only claims to fame are its membership of the zodiac and beautiful M 44.

THE TROPIC OF CANCER

Millennia ago, the Sun reached its summer solstice (its northernmost position in the sky—declination 23.5 degrees north) when it was in front of Cancer. It was then overhead at a northern latitude that was named after this constellation—the Tropic of Cancer. As a result of precession, the Sun's most northerly position has now moved westward to the border of Gemini and Taurus.

The Praesepe or Beehive (M 44)

One of the sky's finest open clusters, this is easy to see through binoculars from the city and with the naked eye from a dark location. There are over 200 stars in the Praesepe. Spread over 1½ degrees, they are best seen with binoculars.

M 67

This open cluster has 500 faint stars spread over half a degree. Although you can find it with binoculars, your best view will be through a small telescope's low-power eyepiece.

R Cancri

This bright long-period variable is easily visible through binoculars when near its 6.2 magnitude maximum. It varies down to 11.2 and back in almost precisely a year.

Stars, I have seen them fall,
But when they drop and die
No star is lost at all
From all the star-sown sky.
STARS, I HAVE SEEN THEM FALL
A. E. HOUSMAN (1859–1936),
English poet

CANCRI (CNC)
On meridian
10 p.m. March 1

KEY

x |

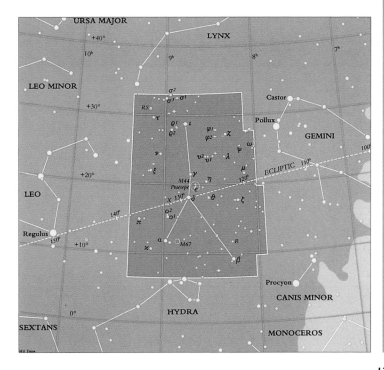

CANES VENATICI (KAH-nez ve-NAT-eh-see)
The Hunting Dogs

This constellation, tucked away just south of the Big Dipper's handle, contains a wide variety of deep sky objects. Conceived by Hevelius in about 1687, Canes Venatici are the hunting dogs, Asterion and Chara, held on a leash by Boötes as he hunts the skies of the north for the bears Ursa Major and Ursa Minor.

Cor Caroli The heart of Charles, Alpha (α) Canum Venaticorum is believed to have been named by Edmond Halley after his patron, Charles II. It is a wide double (separation 20 arc seconds), easily split by the smallest telescope.

M 3 A rare gem of the northern sky, this globular cluster is midway between Cor Caroli and Arcturus. Some 35,000 light years away and 200 light years across, M 3 begins to resolve into stars through a small telescope.

Y Canum Venaticorum (E-B 364) Named La Superba by Secchi in the nineteenth century, this 5th magnitude star is splendidly red. It varies from magnitude 5.2 to 6.6 over 157 days.

The Whirlpool Galaxy (M 51) This famous galaxy appears as a round, 8th magnitude glow with a bright nucleus. A 12 inch (300 mm) telescope will show its spiral structure.

. . . the perceptible Universe exists as a cluster of clusters, irregularly disposed.

EUREKA,
EDGAR ALLAN POE (1809–49),
American writer

**CANUM VENATICORUM
(CVN)
On meridian
10 p.m. May 1**

KEY

x |

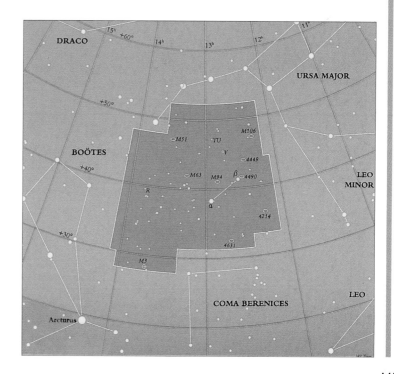

CANIS MAJOR (KAH-niss MAY-jer)
The Great Dog

One of the most striking of all the constellations, the Great Dog is marked by the brilliant star Sirius, commonly known as the Dog Star—the brightest star in the entire sky. Sirius is said to be responsible for the Northern Hemisphere's hot, muggy "dog days" that occur in September.

Sirius The sky's brightest star, Sirius is only 8.7 light years from Earth. Its great brilliance is also due to its being some 40 times more luminous than the Sun. In 1834, Friedrich Bessel noted that Sirius had a strange wobble to its position, indicating an unseen companion. In 1862, the famous telescope maker Alvan Clark, while testing a new 18½ inch (460 mm) refractor on Sirius, discovered the faint star we now know as the Pup. It is a white dwarf star, its density being so great that a piece of it the size of this book might weigh around 200 tons. On its own, the Pup would be a respectable star visible through a telescope at magnitude 8.4, but its closeness to mighty Sirius makes it a difficult target, requiring a telescope of 10 inch (250 mm) aperture and very steady viewing conditions. In ancient Greek and Roman astronomical records, Sirius is quite frequently described as being "ruddy" or "reddish" in color. Was Sirius red in the recent past? Current thinking sees this as unlikely, since other bright stars were sometimes also described as red. This might perhaps have been because of the colors that can be seen when bright stars twinkle.

M 41 A beautiful open cluster, M 41 is surrounded by a rich field of background stars. If you look at it through a telescope, you will be able to see a distinctly red star near the cluster's center.

NGC 2362 This cluster of several dozen stars is tightly packed around Tau (τ) Canis Majoris. What is not clear is whether Tau (τ) is actually a member of the cluster or just a chance foreground star.

SIRIUS AND THE NILE

The ancient Egyptians had a great deal of respect for Sirius. After being close to the Sun for some months, the star would rise just before dawn in late summer, an event known as its heliacal rising. This would herald the annual flooding of the Nile Valley, the waters re-fertilizing the fields with silt. This event was of such importance to the Egyptians that it marked the beginning of their year.

CANIS MAJORIS (CMA)
On meridian
10 p.m. Feb 1

KEY

x 1

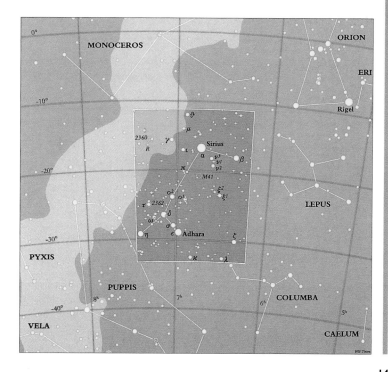

CANIS MINOR (KAH-niss MY-ner)
The Little Dog

Canis Minor has only two stars brighter than 5th magnitude—Procyon (Greek for "before the dog," as it rises before Sirius) and Gomeisa. Besides being one of Orion's hunting dogs, Canis Minor was also said to be one of Actaeon's hounds. One day Actaeon surprised Artemis, goddess of the chase and the forests, while she was bathing in a pond. He paused for a moment and she saw him. Furious that a mortal had seen her naked, Artemis turned him into a stag, set her pack of hounds upon him, and he was devoured.

Nothing puzzles me more than time and space; and yet nothing troubles me less.

LETTER TO THOMAS MANNING,
JANUARY 2, 1810,
CHARLES LAMB (1775–1834),
English essayist

 Procyon (Alpha [α] Canis Minoris) This beautiful deep-yellow star follows Orion across the sky. Only 11.3 light years away, it is accompanied by a white dwarf which is much fainter than the Pup that accompanies Sirius.

Beta (β) Canis Minoris This star is set in a beautiful field which includes one red star.

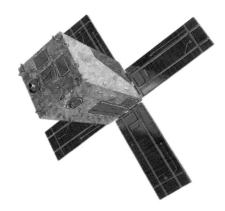

CANIS MINORIS (CMI)
On meridian
10 p.m. Feb 10

KEY

x 1

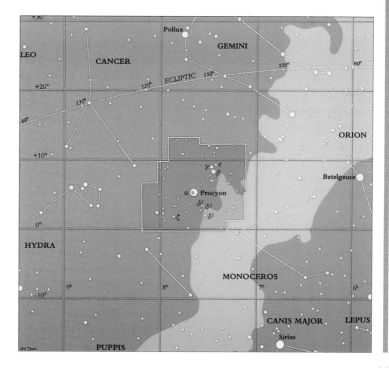

CAPRICORNIS (kap-reh-KOR-nuss)
The Sea Goat, Capricorn

Capricornus has been named for a goat since the time of the Chaldeans and Babylonians. Sometimes it is shown as a goat, but more commonly it is depicted as a goat with the tail of a fish. This might relate to a story about the god Pan, who, when fleeing the monster Typhon, leaped into the River Nile. The part of him that was underwater turned into a fish tail, while his top half remained that of a goat. Capricornus, for northern observers perhaps the zodiac's least visible constellation, can be found by joining Aquila's three brightest stars in a line southward.

👁 Alpha (α) Capricorni

This double has a separation of 6 arc minutes—a naked-eye test for a night's clarity and steadiness. The pair is a double by coincidence, but each star is itself a true binary.

🔭 M 30 Perhaps 40,000 light years away, this globular cluster has a fairly dense center. It is not well resolved in small telescopes.

CAPRICORN

THE TROPIC OF CAPRICORN
Several thousand years ago, the Sun reached its southernmost position in the sky (its winter solstice—declination 23.5 degrees south) when it was in front of Capricornus. During this time it was overhead at a southerly latitude we call the Tropic of Capricorn. It still carries this name, although the Sun, as a result of precession, is now in Sagittarius at the time of the winter solstice.

CAPRICORNI (CAP)
On meridian
10 p.m. Sept 1

KEY

③

x 1.5

EQUULEUS

AQUILA

0°

AQUARIUS

-10°

340°

λ

α² α¹ ξ¹
ν ξ²
τ β

μ

330°

320° ι

δ γ

310° ρ ν ο π σ

-20°

κ ε

φ η

RT 300°

ECLIPTIC 290°

M30 ζ

χ

24

ψ

ω

PISCIS AUSTRINUS

-30°

Fomalhaut

23ʰ 22ʰ 21ʰ 20ʰ SAGITTARIUS

MICROSCOPIUM

CORONA
AUSTRALIS

GRUS

-40° 19ʰ

147

CARINA (ka-RYE-nah)
The Keel

This Southern Hemisphere constellation is in the middle of one of the richest parts of the Milky Way and under a dark sky it is breathtaking. With binoculars, you can see at least half a dozen bright open clusters. Carina is part of what was once a huge constellation known as Argo Navis, the Ship Argo—the vessel in which Jason and his Argonauts sailed on their search for the Golden Fleece. Argo Navis covered such a rich area of sky that it was divided into four separate constellations: Pyxis, Puppis, Vela and Carina.

Canopus (Alpha [α] Carinae) This yellow supergiant is the second brightest star in the sky and is some 74 light years away.

Eta (η) Carinae In 1677, Edmond Halley noticed that this star had brightened. In 1827 it shot up to 1st magnitude, and for a few weeks in 1843 it tied with Sirius as the brightest star in the sky. In recent years, however, Eta (η) has been too faint to be seen without binoculars. The star is famous primarily for the surrounding Eta (η) Carinae Nebula (NGC 3372), the most exquisite nebula in the Milky Way. It is 2 degrees across, with dark rifts appearing to break it up. Superimposed on the brightest part of the nebula is the dark Keyhole Nebula (NGC 3324).

NGC 3532 A brilliant open cluster some 3 degrees from Eta Carinae, this is the finest of the clusters in Carina, with about 150 stars visible in a telescope at low magnification.

IC 2602 This open cluster of scattered, bright stars around Theta (ϑ) Carinae is seen to best advantage in binoculars or in the eyepiece of a wide-field telescope.

CARINAE (CAR)
On meridian
10 p.m. March 1

KEY

②

🖐
x 2

☆

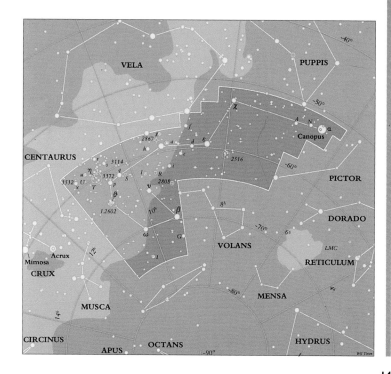

CASSIOPEIA (kass-ee-oh-PEE-uh)
The Queen

This striking W-shaped figure is on the other side of Polaris from the Big Dipper. Most prominent in the Northern Hemisphere's winter sky, Cassiopeia is visible all year from mid-northern latitudes. In Greek mythology, she was queen of the ancient kingdom of Æthiopia—wife of Cepheus and mother of Andromeda. The Romans saw Cassiopeia as having been chained to her throne, as a punishment for her boastfulness, and placed in the heavens to sometimes hang upside down. Arab cultures pictured the constellation as a kneeling camel.

👁 Gamma (γ) Cassiopeiae

This star lies at the center of Cassiopeia's W figure. Normally the constellation's third brightest star, it is an irregular variable. Over a few weeks in 1937, it was the brightest star in the constellation and almost as bright as Deneb in Cygnus. Known as a shell star, Gamma (γ) Cassiopeiae is slowly losing mass into a disk or shell that surrounds it, and alterations in the shell's thickness might be responsible for its variations in brightness.

M 52

This group of about 100 stars is one of the richest in the northern half of the sky, but only one of several open clusters scattered throughout Cassiopeia.

NGC 663

A small open cluster of quite faint stars, NGC 663 is an attractive sight in a small telescope.

. . . Torrent of light and river of the air,
Along whose bed the glimmering stars are seen,
Like gold and silver sands in some ravine . . .

HENRY WADSWORTH LONGFELLOW
(1807–82),
American poet

CASSIOPEIAE (CAS)
On meridian
10 p.m. Nov 1

KEY

x 2

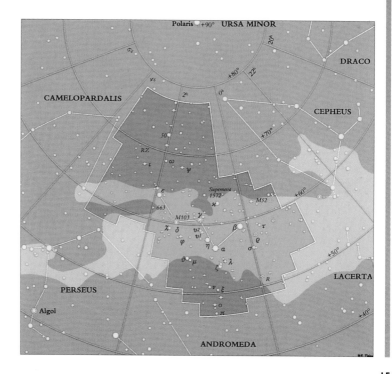

CENTAURUS (sen-TOR-us)
The Centaur

These stars represent Chiron, one of the Centaurs—creatures that were half man, half horse. Chiron was extremely wise, and tutored such humans as Jason and Hercules. Hercules accidentally wounded him, and Chiron, in great pain but unable to die because he was immortal, pleaded with the gods to end his suffering. Zeus mercifully allowed Chiron to die, and placed him among the stars.

Alpha (α) Centauri At the foot of the Centaur, this star is only 4.3 light years away and is the Sun's nearest neighbor. One of the prettiest binary stars, its two components revolve around each other once every 80 years. Alpha (α) and Beta (β) Centauri are the bright "pointers" to the Southern Cross.

Proxima Centauri This faint, 10.7 magnitude star is moving at the same rate and in the same direction as the two stars of Alpha (α) Centauri but about 2 degrees away. A very small red dwarf star only 40,000 miles (25,000 km) across, this star is actually a little closer to us than the other two, but it is presumed to be their companion. It flares occasionally, jumping by half a magnitude or more, usually returning to its normal brightness within half an hour.

Omega (ω) Centauri This globular cluster is perhaps the finest example in the entire sky, with perhaps a million members. Only 17,000 light years away, this is one of the closest clusters to us, second only to NGC 6397 in Ara. Unlike most globulars, it is oval rather than round in shape. A 3 inch (75 mm) telescope will show a large, fuzzy disk with mottled edges, while a 6 inch (150 mm) one will resolve it into stars. Viewed in an even larger telescope, under a dark sky, it looks magnificent, with the field of a low-power eyepiece overflowing with faint stars.

NGC 5128 Located only 4½ degrees north of Omega (ω) Centauri, this elliptical galaxy is distinguished by a strange dark band that crosses its center—probably the result of a collision with a spiral galaxy. A strong source of radio energy, known as Centaurus A, this galaxy emits more than 1,000 times the radio energy of our own galaxy. The dark dust lane is apparent in dark skies with a 4 inch (100 mm) or larger telescope.

NGC 3918 This is a planetary nebula not far from the Southern Cross, presenting a classic blue-green disk about 12 arc minutes across.

CENTAURI (CEN)
On meridian
10 p.m. May 10

KEY

x 2.5

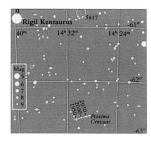

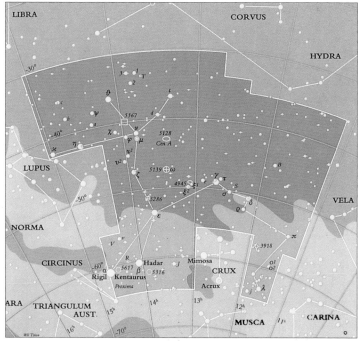

153

CEPHEUS (SEE-fee-us)
The King

Cepheus is an inconspicuous constellation. Its five bright stars are easy to find, only because they face the open side of the W shape of Cassiopeia. It looks a little like a house with a pointed roof. Although the top of the roof does not really point to Polaris, it offers the general direction to the pole at a time of year when the pointer stars of the Big Dipper are not readily accessible. In Greek mythology, Cepheus was King of the ancient land of Æthiopia, husband of Cassiopeia and father of Andromeda.

👁 Delta (δ) Cephei

One of the most famous of the variable stars, and the prototype for the Cepheid variables, Delta (δ) Cephei's variation was discovered by John Goodricke, a deaf-mute teenager, in 1784. Its highest magnitude is 3.5, as bright as neighboring Zeta (ζ) Cephei, and it fades to 4.4, the brightness of Epsilon (ε) Cephei. It completes a cycle every 5.4 days.

👁 Mu (μ) Cephei

This star is so strikingly red that William Herschel called it the Garnet Star. Using Zeta (ζ) and Epsilon (ε) as comparison stars, you can watch it vary in brightness irregularly over hundreds of days.

The stars I know and recognize and even call by name. They are my names, of course. I don't know what others call the stars.

OLD WOMAN QUOTED BY
ROBERT COLES IN
The Old Ones of New Mexico

SKYWATCHING TIP

The weather will determine what you will see on any given night. Moisture of any kind in the air degrades the transparency of the sky. Clear nights usually come after a cold front sweeps out haze and humidity, then brings a dry high-pressure center. Nights with a low dewpoint also tend to be transparent. The dewpoint is the temperature at which moisture in the air condenses, and a low dewpoint means moisture is unlikely to condense out of the atmosphere as the temperature falls during the night.

CEPHEI (CEP)
On meridian
10 p.m. Oct 1

KEY

×2

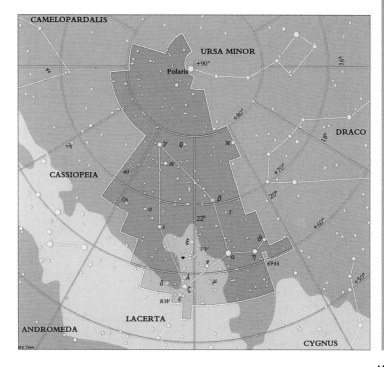

CETUS (SEE-tus)
The Sea Monster

Known by the ancient Greeks as the monster that was about to attack Andromeda when Perseus destroyed it, Cetus was thought in early Christian times to represent the whale that swallowed Jonah. Cetus consists of faint stars, but it occupies a large area of sky. His head is a group of stars not far from Taurus and Aries, and his body and tail lie towards Aquarius.

Mira Omicron (o) Ceti, known as Mira, is the most famous long-period variable. On August 13, 1596, David Fabricius, a Dutch skywatcher, noticed a new star in Cetus. Over the following weeks, it faded, disappeared, then reappeared in 1609. In 1662 Johannes Hevelius named it Mira Stella, the Wonderful Star. Mira varies from a magnitude of about 3.4 to a minimum of about 9.3 over 11 months.

M 77 The brightest of several galaxies in Cetus, M 77 is a 9th magnitude spiral galaxy with a bright core. A 4 inch (100 mm) telescope shows a faint circular disk around the core.

He, who through vast
   immensity can pierce,
See worlds on worlds comprise
   one universe,
Observe how system into
   system runs,
What other planets circle other
   suns, . . .
May tell why Heaven has made
   us as we are.

AN ESSAY ON MAN,
ALEXANDER POPE (1688–1744),
English poet

CETI (CET)
On meridian
10 p.m. Nov 10

KEY

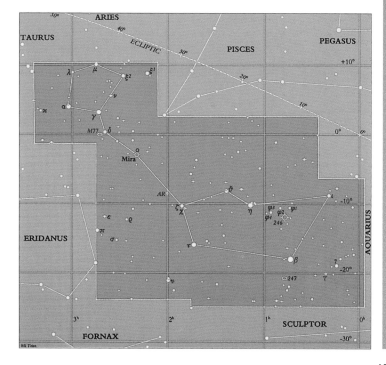

Chamaeleon (ka-MEE-lee-un)

The Chameleon

Johann Bayer drew this constellation early in the seventeenth century, following descriptions that had been given by certain early south sea explorers. The chameleon is a small lizard found in Africa that can change color to match its surroundings. One of the smallest and least conspicuous of the constellations, Chamaeleon does a good job hiding in the sky too. Consisting of a few faint stars, it lies close to the south celestial pole, south of Carina and right beside the south polar constellation of Octans.

SKYWATCHING TIP

When using your telescope, keep in mind that objects that have large diameters and faint magnitudes—as many galaxies do—require low magnification and attention to contrast distinctions, while objects that have small diameters and bright magnitudes—such as planetary nebulae—demand greater magnification and accurate positioning of the telescope's field of view.

Z Chamaeleontis This faint variable star erupts periodically. At its minimum it shines at magnitude 16.2, invisible except in 12 inch (300 mm) or larger telescopes. However, every three to four months it undergoes an outburst, rising within a few hours to about magnitude 11.5, and for a few days it is visible through a 6 inch (150 mm) telescope. Even so, this does not constitute an easy target in a corner of the sky with few stars.

The Universe is an infinite sphere, the center of which is everywhere, the circumference nowhere.

Pensées,
Blaise Pascal (1623–62),
French mathematician and
natural philosopher

CHAMAELEONTIS (CHA)
On meridian
10 p.m. April 1

KEY

x 1

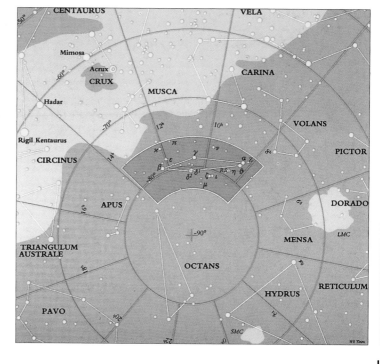

CIRCINUS (SUR-seh-nus)
The Drawing Compass

The early explorers in the south seas were less interested in mythology than in the modern instruments that they relied on to find their way around uncharted waters. The Drawing Compass is one of a number of obscure constellations that were designated by the French astronomer Nicolas-Louis de Lacaille. He worked at an observatory at the Cape of Good Hope from 1750 to 1754, where he compiled a catalogue of more than 10,000 stars.

Alpha (α) Circini At only 3rd magnitude, this is the constellation's brightest star. It lies just near the much brighter Alpha (α) Centauri. It is about 65 light years away and has a faint 9th magnitude companion.

SKYWATCHING TIP
Scale is one of the hardest things to become accustomed to when observing the night sky. If, in your early attempts, you are unable to find a single constellation it might be because you have no idea what size pattern you are looking for. Firstly, look at the key to the constellation chart to see how many hand spans will cover the constellation from east to west. Then compare the new constellation chart with some familiar ones.

CIRCINI (CIR)
On meridian
10 p.m. June 1

KEY

x 1

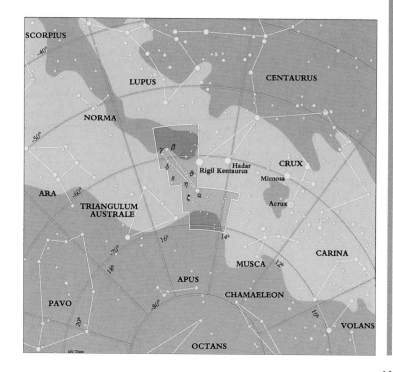

COLUMBA (koh-LUM-bah)
The Dove

Immediately south of Canis Major, Columba is a modern constellation named by Petrus Plancius, a sixteenth-century Dutch theologian and mapmaker. This inconspicuous group of stars honors the dove that Noah sent out from the ark after the rains had stopped, to see if it could find dry land.

How beautiful is night!
A dewy freshness fills the
  silent air;
No mist obscures, nor cloud,
  nor speck, nor stain,
Breaks the serene of heaven.

THALABA THE DESTROYER
ROBERT SOUTHEY
(1774–1843),
English poet

T Columbae A Mira variable, this star has a maximum magnitude of 6.7. It drops to magnitude 12.6 and then rises again over a period of seven and a half months.

NGC 1851 Bright and large, this 7th magnitude globular cluster appears as a misty spot through binoculars under a good sky. A 6 inch (150 mm) telescope will begin to resolve the cluster's brightest stars.

COLUMBAE (COL)
On meridian
10 p.m. Jan 20

KEY

x 1

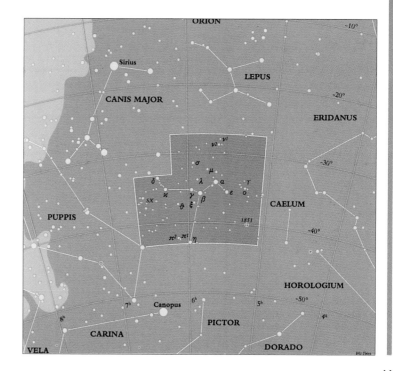

COMA BERENICES (KOH-mah bear-eh-NEE-seez)
Berenice's Hair

Between Arcturus and Denebola (Beta [β] Leonis), Coma Berenices has no bright stars and is hard to distinguish, but it is a remarkable area of sky. It is a sprinkling of faint stars superimposed on a cloud of galaxies—the northern end of the Virgo cluster of galaxies. The constellation's name commemorates Berenice, the beautiful wife of the ancient Egyptian king Ptolemy III, who sacrificed her long golden hair to Aphrodite in return for her husband's safe return from battle.

Someday I would like to stand on the moon, look down through a quarter of a million miles of space and say, "There certainly is a beautiful earth out tonight."
LT. COL. WILLIAM H. RANKIN
(b. 1920),
American author

M 53 This fine globular cluster is about 3 arc minutes in diameter and is located close to Alpha (α) Comae Berenices.

The Blackeye Galaxy (M 64) The Blackeye is one of the most unusual galaxies in the sky. It looks like an ordinary spiral galaxy, with tightly wound arms, but viewed with a 4 to 6 inch (100 to 150 mm) telescope or larger, a huge cloud of dust can be seen dominating its center, giving it the look of a black eye.

NGC 4565 Under a dark sky, a small telescope should show this faint object as a pencil-thin line of haze. It is a spiral galaxy seen edge-on, with a dust lane that becomes apparent in 8 inch (200 mm) telescopes.

COMAE BERENICES (COM)
On meridian
10 p.m. May 1

KEY

x I

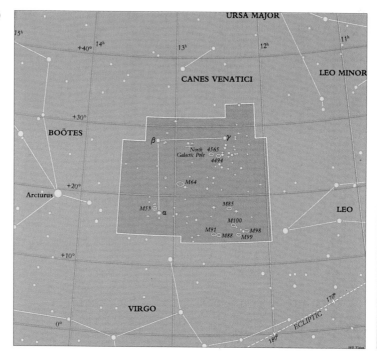

CORONA AUSTRALIS (kor-OH-nah os-TRAH-lis)
The Southern Crown

One of the 48 original constellations catalogued by the Egyptian astronomer Ptolemy in the second century AD, this small semicircular group of faint stars is inconspicuous, especially from the Northern Hemisphere. It lies just south of Sagittarius and is said to represent a crown of laurel or olive leaves, as given to victorious athletes in ancient times.

NGC 6541 This globular cluster presents a small nebulous disk to smaller telescopes. An 8 inch (200 mm) telescope only begins to resolve the edge into stars.

Whereas other animals hang their heads and look at the ground, he made man stand erect, bidding him look up to heaven, and lift his head to the stars.

METAMORPHOSIS,
OVID,
(43 BC–AD 17?),
Roman poet

CORONAE AUSTRALIS
(CRA)
On meridian
10 p.m. Aug 1

KEY

x 1

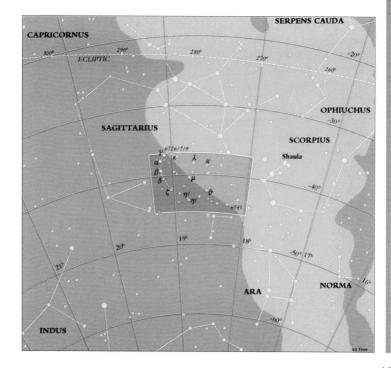

CORONA BOREALIS (kor-OH-nah bor-ee-AL-is)
The Northern Crown

Just 20 degrees northeast of Arcturus lies the Northern Crown, a small semicircle of stars that are faint but very distinct. Greek myth claims the crown belongs to Ariadne, daughter of Minos, King of Crete. Ariadne was reluctant to accept a marriage proposal from Dionysus (who was in mortal form), since she did not wish to marry a mortal. To prove he was a god, Dionysus took off his crown and threw it into the heavens as a tribute to her. Satisfied, Ariadne married him and became immortal herself.

R Coronae Borealis One of the more remarkable stars in the sky, R Cor Bor, as it is usually known, is a nova in reverse. Normally shining at magnitude 5.9, at completely irregular intervals the star will suddenly fade, sometimes by as much as 8 magnitudes, as dark material erupts in its atmosphere. It then slowly recovers as the material dissipates.

T Coronae Borealis Now shining at magnitude 10.2, in 1866 this star suddenly rose to magnitude 2. Known as a recurrent nova, the star repeated the performance unexpectedly in 1946, and will probably do so again.

We shall not cease from exploration, and the end of all our exploring will be to arrive where we started and know the place for the first time.

LITTLE GIDDING,
T.S. ELIOT (1888–1965),
British poet, dramatist and critic

CORONAE BOREALIS
(CRB)
On meridian
10 p.m. June 30

KEY

x 0.5

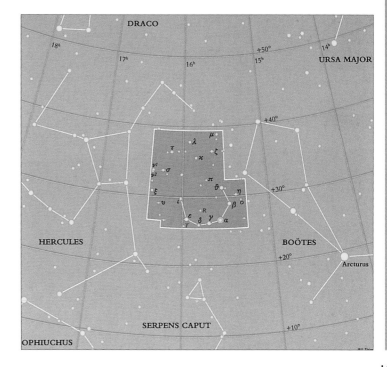

CORVUS (KOR-vus) & CRATER (KRAY-ter)
The Crow & the Cup

Sent one day by Apollo for a cup of water, Corvus was slow in returning as he had been waiting for a fig near the spring to ripen. Bringing the cup (Crater) of spring water and a water serpent (Hydra) back in his claws, he told Apollo that he had been delayed because the serpent had attacked him. Apollo, knowing Corvus was lying, placed all three in the sky. The Cup is to the west of Corvus, within reach, but the serpent prevents him from drinking from it.

R Corvi This Mira-type variable star ranges from magnitude 6.7 to 14.4 over a period of about 10 months.

Tombaugh's Star This very faint cataclysmic variable star, TV Corvi, was discovered as a nova by Clyde Tombaugh in 1931, while searching for planets.

The Ring-tailed Galaxy Also called the Antennae or Rat-tailed Galaxy, NGC 4038 and NGC 4039 form a faint, 11th magnitude pair of galaxies that are interacting or colliding. Needing an 8 inch (200 mm) telescope to see, it is still one of the brightest pairs of connected galaxies.

FINDING CORVUS

Arc to Arcturus, speed to Spica, then turn west and you will see Corvus—a small foursome of stars. Crater is a fainter constellation alongside that looks like a cup.

At length, by sparing neither labor nor expense, I succeeded in constructing for myself an instrument so superior that objects seen through it appear magnified nearly a thousand times...

GALILEO GALILEI (1564–1642), Italian astronomer, mathematician and physicist

CORVI (CRV)
CRATERIS (CRT)
On meridian
10 p.m. April 20

KEY CRV	KEY CRT
③	④
x l	x l

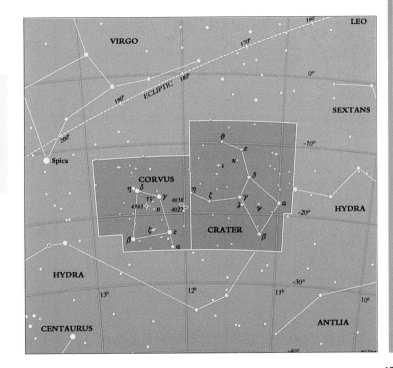

CRUX (KRUKS)
The Southern Cross

The most famous southern constellation, the Southern Cross appears on the flags of several nations. Its distinctive pattern of stars helped guide sailors for centuries, the upright of the cross pointing the way to the south celestial pole. Because it lies so far south, Crux was not mapped as a separate entity until 1592. Until then, it formed part of Centaurus. The cross contains the most striking pair of opposites—the brightness of the Jewel Box and the dark of the Coal Sack—embedded within the southern Milky Way.

Acrux This is the popular name for Alpha (α) Crucis, the bright double star at the foot of the cross, separated by approximately 4½ arc seconds. A third star, quite bright at magnitude 5, lies 90 arc seconds away.

Gamma (γ) Crucis Also known as Gacrux, this wide double star marks the northern end of the Southern Cross. An optical double, it consists of a magnitude 6.4 star lying almost 2 arc minutes from a bright orange primary.

The Jewel Box Superimposed on Kappa (κ) Crucis, this is one of the finest open clusters. Although small, it sparkles in any instrument, and has several stars of contrasting color.

The Coal Sack This is one of the largest and densest dark nebulae in the sky. It lies just east of Acrux and is clearly visible in a dark sky against the star clouds of the Milky Way.

Time and space are fragments of the infinite for the use of finite creatures.

JOURNAL, NOVEMBER 16, 1824,
HENRI FRÉDÉRIC AMIEL (1821–81),
French writer

CRUCIS (CRU)
On meridian
10 p.m. May 1

KEY

× 0.5

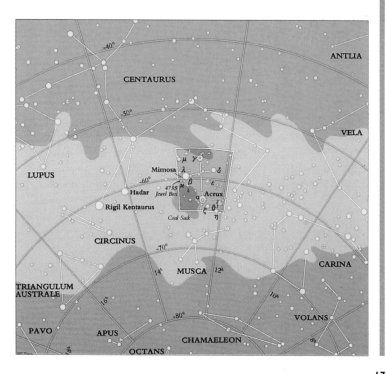

Sky and Constellation Charts

173

CYGNUS (SIG-nus)
The Swan

Cygnus is the Northern Hemisphere's answer to Crux. Looking like a large cross, Cygnus straddles the northern Milky Way. If you are under a dark sky you may be able to see the Milky Way divide into two streams in Cygnus. A dark nebula between us and the more distant stars causes this apparent divergence.

👁 Deneb (Alpha [α] Cygni)

Deneb means "tail" in Arabic, which is where this star is positioned on the swan. On a par with Rigel in Orion, it is one of the mightiest stars known—25 times more massive and 60,000 times more luminous than the Sun. About 1,500 light years away, Deneb is by far the most distant star of the famous Summer Triangle, which it forms with Vega and Altair. Vega is 25 light years away and Altair only 16.

👁 Albireo (Beta [β] Cygni)

Albireo, at the foot of the cross, is one of the prettiest sights in the sky. Without a telescope it is seen as a single star; a telescope transforms it into a spectacular double with a separation of 34 arc seconds. One member is golden yellow with a magnitude of 3, and the other is blueish with a magnitude of 5.

🔭 61 Cygni

Dubbed the Flying Star because of its rapid motion relative to more distant stars, this double is easily separated in small telescopes. The two components revolve around each other over the course of about 650 years.

👁 The North America Nebula (NGC 7000)

This giant cloud is illuminated by Deneb, which lies only 3 degrees to the west. Because of its size, the nebula is difficult to see in a telescope: it is best seen with the naked eye on a dark night. Photographs show this nebula to look surprisingly like the shape of North America, but this resemblance is not readily apparent to the eye when observing.

M 39

This loosely bound open star cluster is seen at its best through binoculars. On a clear night you might be able to see it with the naked eye.

Chi (ψ) Cygni

At maximum brightness, typically magnitude 4 or 5, this long-period variable is bright enough to be seen with the naked eye. It fades to magnitude 13 and then climbs back in a period of a little more than 13 months.

🔭 The Veil Nebula (NGC 6960, 6992, 6995)

The lacy remnants of a supernova, this

CYGNI (CYG)
On meridian
10 p.m. Aug 20

KEY

①

✋

x 1.5

☆

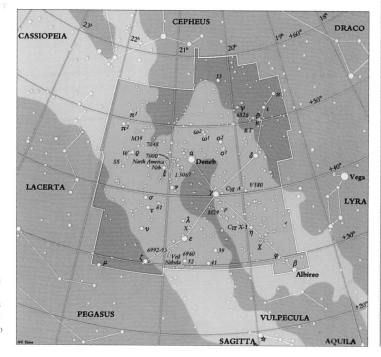

beautiful nebulosity requires at least a 6 inch (150 mm) telescope. NGC 6960, the nebula's western arc, passes through 52 Cygni, which makes it easier to find but harder to see.

DELPHINUS (del-FIE-nus)
The Dolphin

A small constellation with a distinctive shape, Delphinus has been thought of as a dolphin since ancient times. It is said that the mermaid Amphitrite agreed to marry Poseidon from whom she had been trying to escape, on the advice of a dolphin. Poseidon, the god of the sea, was so pleased with the little dolphin that he placed him among the stars. The constellation is also sometimes known as Job's Coffin, the origin of which is obscure.

Gamma (γ) Delphini This is an optical double with a separation of 10 arc seconds. The brighter is magnitude 4.5 and the fainter, which is slightly green, is 5.5.

R Delphini This Mira star has a magnitude range of 8.3 to 13.3 over a period of 285 days.

NICOLO CACCIATORE

Delphinus's alpha (α) star is named Sualocin and its beta (β) star is known as Rotanev. These names honor a relatively recent observer, Niccolo Cacciatore, long-time associate of the famous nineteenth-century observer Giuseppe Piazzi. Star atlases at the time included these names without comment, but the Reverend Thomas Webb worked out that the names, spelled backward, are Nicolaus Venator—the Latinized version of Cacciatore's name.

DELPHINI (DEL)
On meridian
10 p.m. Sept 1

KEY

x I

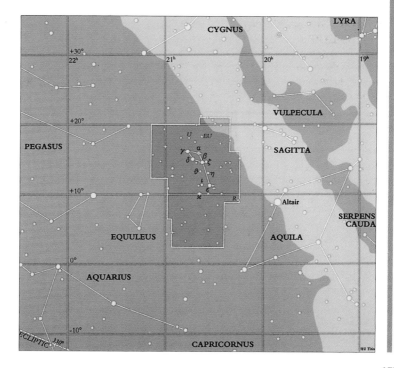

177

DORADO (doh-RAH-doh)
The Dolphinfish

Far to the south, this constellation was first recorded by Bayer in his star atlas of 1603. Dorado does not honor the tiny fish in many a home aquarium, but the tropical dolphinfish, the mahi-mahi, member of the Coryphaenidae family, which can be more than 5 feet (1.75 m) long. Since they swim fast and often leap out of the water in play, sailors used to consider their appearance a good omen.

Light from distant places has made the journey to earth, and it falls on these new eyes of ours, the telescopes...

COSMIC LANDSCAPE,
MICHAEL ROWAN-ROBINSON
(b.1942),
British mathematician

 The Large Magellanic Cloud (LMC) This is a companion galaxy to the Milky Way, lying 168,000 light years away—less than one-tenth the distance to the Andromeda Galaxy (M 31). As a result, it spans about 11 degrees of the sky, presenting its contents to the scrutiny of Southern Hemisphere observers. It was from this galaxy that supernova 1987A blazed forth. The LMC is plainly visible in a dark sky, but it is easily lost in the glare of city lights.

The Tarantula Nebula (NGC 2070) Also known as the 30 Doradus Nebula, this is one of the finest emission nebulae in the sky, despite its distance from us. Its true size is perhaps 30 times that of the more famous Great Nebula in Orion (M 42).

 S Doradus A single super-luminous star within the open cluster NGC 1910, S Doradus varies irregularly in brightness between magnitudes 8 and 11. It is one of the most luminous stars known.

DORADUS (DOR)
On meridian
10 p.m. Jan 1

KEY

③

🖐
x 1

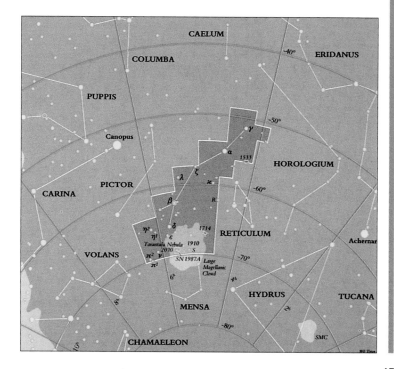

179

DRACO (DRAY-koh)
The Dragon

This constellation is circumpolar from much of the Northern Hemisphere and is best seen during the warmer months. A large, faint constellation, the Dragon is hard to trace as it winds about between Ursa Major, Boötes, Hercules, Lyra, Cygnus and Cepheus. The Chaldeans, Greeks and Romans all saw a dragon here, while Hindu mythology claims the creature is an alligator. The Persians saw a man-eating serpent. Thuban, the brightest star in the constellation, was the pole star in ancient times, but the Earth's precession has since moved the pole to Polaris.

👁 **Quadrantids** This is one of the strongest meteor showers. The time of maximum activity is around January 3, although the shower lasts only a few hours.

👁 **Draconids** This meteor shower consists of particles from Periodic Comet Giacobini-Zinñer. In 1933 and 1946, the shower's date of October 9 closely followed the comet's crossing of the Earth's orbit, and the result was a spectacular storm of meteors.

🔭 **NGC 6543** This 8th magnitude planetary nebula lies midway between the stars Delta (δ) and Zeta (ζ) Draconis. It is bright blue-green in color, but high power is needed in order to make out its small, hazy disk.

DRACONIS (DRA)
On meridian
10 p.m. July 1

KEY

③

× 2.5

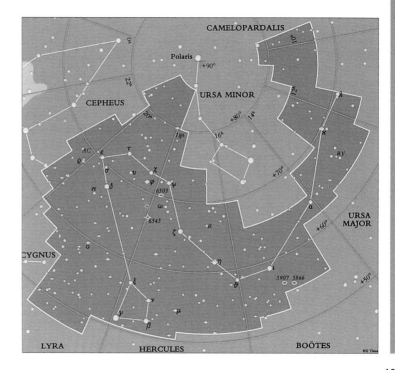

EQUULEUS (eh-KWOO-lee-us)
The Little Horse

With the exception of Crux, Equuleus occupies a smaller patch of sky than any other constellation. It lies just to the southeast of Delphinus and because it has no bright stars it is of limited interest. The famous Greek astronomer Hipparchus is thought to have made up the constellation in the second century BC. It has been said to represent Celeris, brother of Pegasus (the Winged Horse), given to Castor (one of the twins represented by Gemini) by Mercury.

As I flit through you hastily,
soon to fall and be gone,
what is this chant,
What am I myself but one of
your meteors?

YEAR OF METEORS,
WALT WHITMAN (1819–92),
American poet

SKYWATCHING TIP

When skywatching, it's a good idea to keep a record of what you see. Not only will this help you to remember your observations, but the effort to record details will stimulate careful and meaningful viewing. A notebook that's large enough for drawings as well as notes is ideal. The information you take down should include: date, time and location of observation; instrument(s) used; seeing conditions; and sketches of your sightings.

Alpha (α) Equulei The constellation's brightest star is named Kitalpha, which is Arabic for "little horse."

EQUULEI (EQU)
On meridian
10 p.m. Sept 1

KEY

④

×0.5

ERIDANUS (eh-RID-an-us)
The River

This constellation has been seen as a river since ancient times—usually the Euphrates or the Nile. For early observers in South-west Asia, the river extended only as far south as Acamar, or Theta (ϑ) Eridani, because they could not see the stars that lay further south. In Book II of *Metamorphoses*, Ovid writes of Phaethon being tossed out of the chariot of the Sun to drown in Eridanus.

Omicron 2 (o₂) Eridani

This remarkable triple consists of a 4th magnitude orange dwarf, a 9th magnitude white dwarf, and an 11th magnitude red dwarf. The red and white dwarfs form a pair (separation 8 arc seconds), and are separated from the brighter star by over 80 arc seconds. The white dwarf is the only one of its class that is easy to see in a small telescope.

Epsilon (ϵ) Eridani

Only 10.8 light years away, this star is a smaller version of our Sun. Radio telescopes have searched but unsuccessful in discovering signals indicating intelligent life.

THE RIVER'S COURSE

What a long constellation the river Eridanus is! Its source lies immediately to the west of Rigel in Orion, with a star called Cursa or Beta (β) Eridani. It flows southward until it reaches its mouth in Achernar (Alpha [α] Eridani) near the south celestial pole—a very bright star few northern observers ever see. From the Southern Hemisphere, an observer can follow the full course of the river, even though the stars are faint.

ERIDANI (ERI)
On meridian
10 p.m. Dec 10

KEY

×2

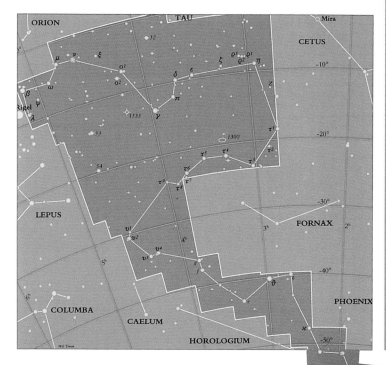

FORNAX (FOR-nax)
The Furnace

When Nicolas-Louis de Lacaille invented this constellation out of several faint stars in a bend of River Eridanus, he named it Fornax Chemica, the Chemical Furnace, in honor of the chemist Antoine Lavoisier, who was guillotined during the French Revolution in 1794. Today it is simply known as the Furnace.

SKYWATCHING TIP

Nebulae, star clusters and galaxies present some of the most interesting targets for amateur astronomers. As a general rule, low magnification and a wide field of view are necessary for deep-sky observing, but don't be afraid to experiment.

The Fornax Galaxy Cluster
While there are no bright points of interest in Fornax, if you have a large telescope you will enjoy this challenging cluster of galaxies near the Fornax–Eridanus border. With a wide-field eyepiece, you may see up to nine galaxies in a single field of view. The brightest galaxy, at 9th magnitude, NGC 1316 is also the radio source Fornax A.

The Fornax System
Although this dwarf galaxy—a member of our Local Group of galaxies—appears to be unusual, galaxies like it may be common in the universe. Spherical in shape, it is a large group of very faint stars that includes a few globular clusters. It is too faint to see with an amateur telescope, but one globular cluster, NGC 1049, is, at magnitude 12.9, visible in a 10 inch (250 mm) telescope under a good sky.

KEY

x 1.5

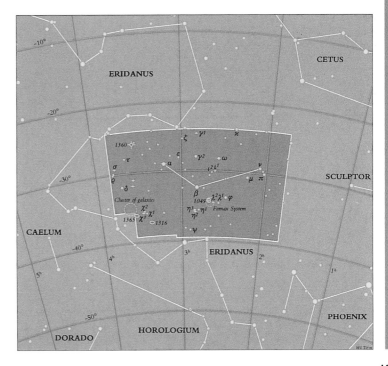

Sky and Constellation Charts

187

GEMINI (JEM-eh-nye)
The Twins

A familiar pattern in the sky, Gemini is part of the zodiac. Various cultures have seen the stars as twins—either as gods, men, animals or plants. The Greeks named the constellation's two brightest stars Castor and Pollux, after the twins who hatched from an egg from their mother Leda, following her seduction by Zeus. The twins were among the heroes who sailed with Jason in the quest for the Golden Fleece. They helped save the *Argo* from sinking during a storm, so the constellation was much valued by sailors. William Herschel discovered Uranus near Eta (η) Geminorum in 1781, and Clyde Tombaugh discovered Pluto near Delta (δ) Geminorum in 1930.

Castor (Alpha [α] Geminorum) This sextuple star can be seen only as a double through a small telescope. Its current separation is about 3 arc seconds.

Eta (η) Geminorum This bright semi-regular variable varies from magnitude 3.2 to 3.9 and back over about eight months.

M 35 This bright open cluster is beautiful through binoculars and spectacular in a small telescope. NGC 2158 is a smaller, fainter open cluster on its southwest edge. It appears in small telescopes as a smudge, being about 16,000 light years away—five times the distance to M 35.

The Clownface or Eskimo Nebula (NGC 2392) This 8th magnitude nebula has a bright central star. The blue-green tint of the nebula's disk gives it away as a planetary type.

You must not expect to see at sight...Seeing is in some respects an art which must be learned.

WILLIAM HERSCHEL (1738–1822), English astronomer

GEMINORUM (GEM)
On meridian
10 p.m. Feb 1

KEY

x 1.5

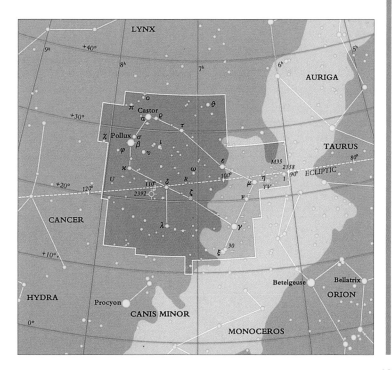

GRUS (GROOS)
The Crane

In his star atlas of 1603, Johann Bayer named this southern constellation Grus, the Crane—the bird which served as the symbol of astronomers in ancient Egypt. This group of stars—variously seen as a stork, a flamingo and a fishing rod—has very little to offer the skywatcher who is using a small telescope, although some faint galaxies provide suitable targets for telescopes of 8 inch (200 mm) aperture or larger. Grus has only three fairly bright stars, which can be used as a simple illustration of magnitude.

...look, how the floor of heaven
Is thick inlaid with patines of
bright gold.

THE MERCHANT OF VENICE,
WILLIAM SHAKESPEARE (1564–1616)

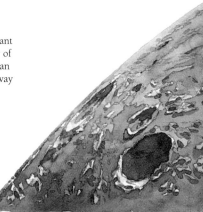

Alpha (α) Gruis Also known as Alnair, this is a large, blue main-sequence star over 100 times as luminous as the Sun. Being 100 light years away, it is the brightest of the three stars only because it is relatively close to us.

Beta (β) Gruis This is a much larger red giant star, several hundred times as luminous as the Sun, but its orange red color makes it appear fainter than Alpha (α) Gruis.

Gamma (γ) Gruis This blue giant star is more luminous than either of the others, and appears fainter than them since it is 200 light years away

KEY

x 1

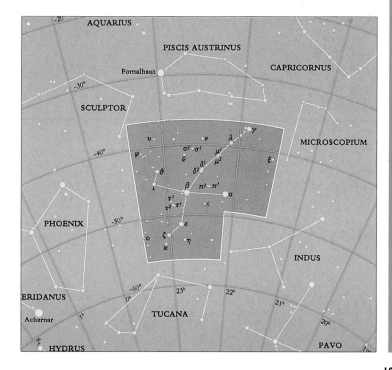

HERCULES (HER-kyu-leez)
Hercules

For northern observers, Hercules, with its "key-stone" of four stars—Epsilon (ϵ), Zeta (ζ), Eta (η) and Pi (π)—is one of the best of the summer constellations. One of the most famous of all the classical heroes, Hercules was immensely strong and was revered throughout the Mediterranean. He was the half-mortal son of Jupiter and was involved in many noble exploits, the most famous being the undertaking of the 12 labors. At the end of his life, as a reward for his bravery, Jupiter made him one of the gods, placing him in the sky.

We are all in the gutter, but some of us are looking at the stars.

OSCAR WILDE (1854–1900),
British playwright,
poet and novelist

The Hercules Cluster (M 13) The most dramatic globular cluster in the northern sky, this is faintly visible to the naked eye as a fuzzy spot, but through a telescope it is a sight to behold. The edges begin to resolve into stars in a 6 inch (150 mm) telescope. When you view this cluster, you are looking 23,000 years into the past.

M 92 M 13's slightly smaller and fainter cousin, this cluster of stars is 26,000 light years away.

Ras Algethi (Alpha [α] Herculis) This is a very red star, varying from magnitude 3.1 to 3.9. It is also a splendid colored double, with a 5th magnitude blue-green companion about 5 arc seconds away from an orange primary.

HERCULIS (HER)
On meridian
10 p.m. July 10

KEY

x 2

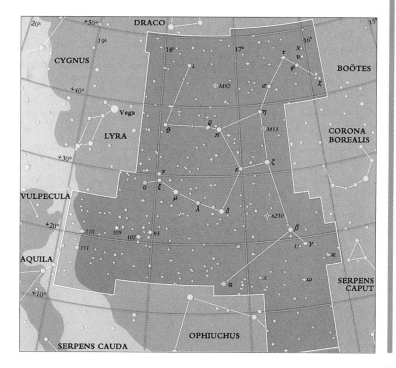

HOROLOGIUM (hor-oh-LOH-jee-um)
The Clock

A small group of stars lying east of Archernar, this is one of the constellations mapped by Nicolas-Louis de Lacaille. Originally called Horologium Oscillatorium, it honors the invention of the pendulum clock by Dutch scientist Christiaan Huygens in 1656 or 1657. Huygens was a leader in Renaissance thinking. By applying the law of the pendulum discovered by Galileo to clockmaking, he significantly increased the accuracy of timekeeping.

The heavens call to you, and circle around you, displaying to you their eternal splendors...

"PURGATORIO" FROM
THE DIVINE COMEDY,
DANTE ALIGHERI (1265–1321),
Italian writer

SKYWATCHING TIP

As you become more involved in observing and develop connections with your local astronomy club, you might be invited to a star party sponsored by the club. Take full advantage of this opportunity to meet people, ask questions, share experiences, compare notes, and look through the various telescopes that are being used. Remember to take a red flashlight with you, so that you can operate effectively in the dark without inconveniencing anyone, and always point it downward, away from people's dark-adapted eyes.

R Horologii This long-period variable star was discovered from an observing station that Harvard University used to run in Peru. In 131/2 months, it completes its cycle of variation from 5th to 14th magnitude and back.

NGC 1261 This 8th magnitude globular cluster is only 6 arc minutes across so is a target for a larger telescope.

HOROLOGII (HOR)
On meridian
10 p.m. Dec 10

KEY

④

🖐

x 1

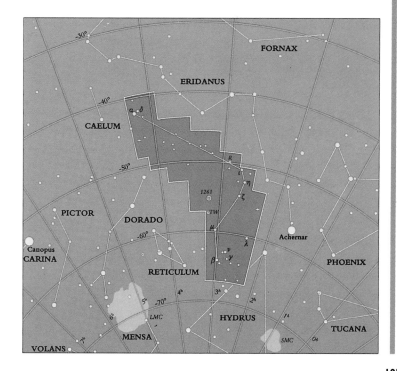

HYDRA (HY-dra)
The Sea Serpent

Hydra was the nine-headed serpent that Hercules had to kill as one of his 12 labors. Each time he lopped off one head, two others grew in its place. Hercules emerged from this nightmare by having his nephew burn the stump of each severed neck, preventing new heads from sprouting.

R Hydrae This Mira star varies over 13 months from a maximum as high as magnitude 3.5 to a minimum of 10.9.

V Hydrae This is a low-temperature red giant is so deeply red that you can be sure you have found it merely by its color. The star varies erratically between magnitudes 6 and 12, with two superimposed periods—one about 18 months, and the other 18 years.

M 48 (NGC 2548) This open cluster is best seen using binoculars or a wide-field telescope.

M 83 This is a spiral galaxy with three obvious spiral arms. At 8th magnitude, it is one of the brighter galaxies visible in binoculars.

The Ghost of Jupiter Nebula (NGC 3242)
This bright planetary nebula is about 16 arc seconds across and shows its structure well in a 10 inch (250 mm) telescope.

HYDRAE (HYA)
On meridian
10 p.m. April 1

KEY

VIRGO

LIBRA

LUPUS

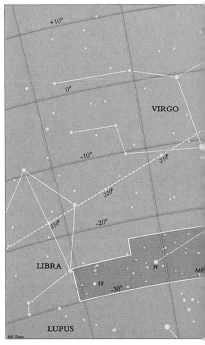

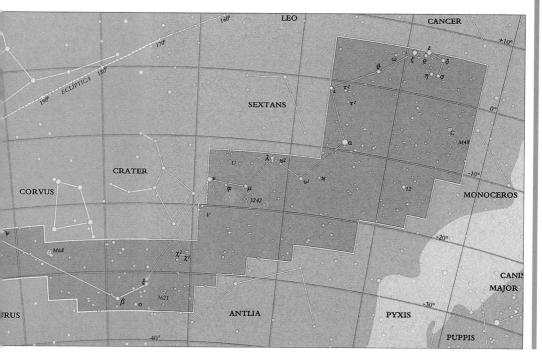

HYDRUS (HY-drus)
The Water Snake

Johann Bayer created this constellation, publishing it in his 1603 star atlas. He placed it near Achernar, the mouth of the River Eridanus, between the Large and Small Magellanic Clouds. It is sometimes called the Male Water Snake, to avoid confusion with Hydra.

VW Hydri This star is the most popular cataclysmic variable with Southern Hemisphere observers. When in its usual state, it shines at a faint 13th magnitude, but when it goes into outburst, an event which occurs about once a month, it can become brighter than 8th magnitude in just a few hours.

Ice is the silent language of the peak;
and fire the silent language of the star.

AND IN THE HUMAN HEART,
CONRAD AIKEN (1889–1973),
American poet

SKYWATCHING TIP

One frigid night in Flagstaff, Arizona, Clyde Tombaugh—the astronomer who discovered Pluto—was staring into the eyepiece of his telescope, guiding a one-hour exposure. Feeling sleepy, he struggled to keep from dozing off. When the hour was up, he realized that he had become so cold that he could hardly move. In great pain, he managed to close the telescope and move to a warm room, where he had to sit for some time beside a heater to thaw out. Be careful not to become so absorbed in your observations that you are unaware of how cold you are. Walk around the telescope to stimulate your circulation, or go inside.

HYDRUS (HYI)
On meridian
10 p.m. Dec 1

KEY

x |

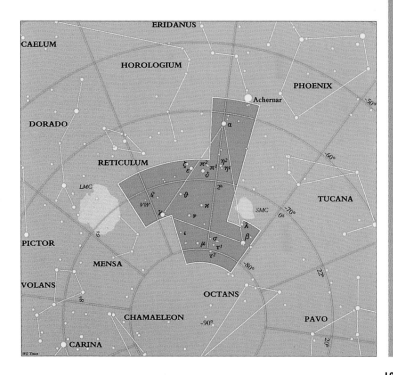

INDUS (IN-dus)
The Indian

This constellation was added to the southern sky by Johann Bayer to honor the Native Americans that European explorers encountered on their travels. The figure of Indus is positioned between three birds: Grus, the Crane; Tucana, the Toucan; and Pavo, the Peacock.

Epsilon (ε) Indi Only 11.3 light years away, this is one of the closest stars to the Sun and is somewhat similar to it. With four-fifths of the Sun's diameter and one-eighth its luminosity, scientists consider Epsilon (ε) Indi to be worth investigating for planets and for evidence of extraterrestrial intelligence, such as radio signals. In the early 1960s, when Frank Drake began searching for signs of life elsewhere in the galaxy, he used this star as one of his targets. In 1972, the Copernicus Satellite searched unsuccessfully for laser signals from this star.

There is in space a small black
* hole*
Through which, say our
* astronomers,*
The whole damn thing, the
* universe,*
Must one day fall. That will
* be all.*

COSMIC COMICS,
HOWARD NEMEROV (1920–91),
American poet

INDI (IND)
On meridian
10 p.m. Sept 1

KEY

x 1.5

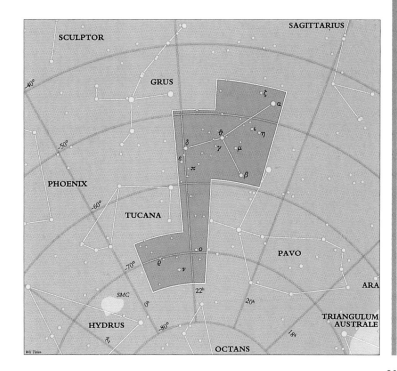

LACERTA (lah-SIR-tah)
The Lizard

Lacerta is far enough north to be circumpolar at the higher mid-northern latitudes. It lies south of Cepheus. The German astronomer Johannes Hevelius suggested that this group of stars be named Lacerta in 1690, but a few revisions were needed before it evolved from a small, long-tailed mammal into a lizard. Other cartographers came up with names for the region to honor France's Louis XIV and Prussia's Frederick the Great, but these names were ignored.

BL Lacertae Since this object varies from 13.0 to 16.1, it is invisible to any but the largest amateur telescopes. However, BL Lacertae is worth a look, since it is not a star at all but the nucleus of a distant elliptical galaxy. Some of this class of BL Lacertae-type (BL Lac) objects have been known to change by as much as two magnitudes in a single day. Recent theories suggest that BL Lac objects, quasars and other high-powered galaxies are all closely related "active galaxies." The powerful energy source at the center may be a black hole surrounded by a complex, swirling mass of gas and dust.

SKYWATCHING TIP

When observing a variable star, set yourself a challenge by trying to estimate the magnitude of the star. This can be done by using nearby stars for comparison. For example, if Delta (δ) Cephei (see p.154) is a little fainter than the magnitude 3.5 star Zeta (ζ) and much brighter than the 4.4 Epsilon (ε), then Delta (δ) Cephei will be about magnitude 3.6 or 3.7. But what if you do not know the magnitudes of the comparison stars, a and b? If the variable, V, were 3/4 of the way from a to b in brightness, you could write your estimate as "a, 3,V, 1, b". This way you can monitor the changing brightness of V over time.

LACERTAE (LAC)
On meridian
10 p.m. Oct 1

KEY

× 0.5

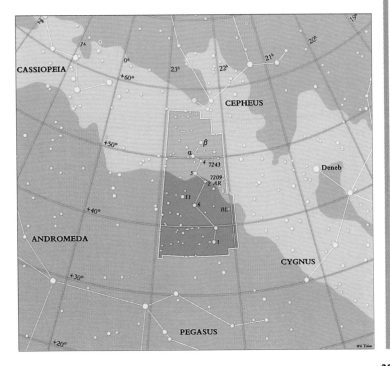

LEO (LEE-oh) & LEO MINOR (LEE-oh MY-ner)
The Lion & the Little Lion

Unlike most of the zodiacal constellations, Leo, with its sickle (or backward question mark) tracing out a great head, really can be pictured as its namesake, a lion, often shown reclining not unlike the Egyptian Sphinx. The Babylonians and other cultures of Southwest Asia associated Leo with the Sun, because the summer solstice occurred when the Sun was in that part of the sky. Leo Minor is a recent addition to the constellations, introduced by Johannes Hevelius during the seventeenth century.

Gamma (γ) Leonis This beautiful double star has orange-yellow components of 2nd and 3rd magnitude separated by 5 arc seconds.

R Leonis This favorite Mira variable is easily found near Regulus. It ranges from magnitude 5.9 to 11 over about 10½ months.

R Leonis Minoris Another Mira star, this one takes about a year to vary between magnitudes 7.1 and 12.6.

M 65 and M 66 These two spiral galaxies near Theta (ϑ) Leonis are visible in binoculars but give a better view in a telescope. Other interesting galaxies in Leo are NGC 3628, M 95, M 96, M 105 and NGC 2903.

Leonids This meteor shower peaks annually on November 17. In 1966, observers recorded up to 40 meteors per second at its peak.

LEONIS (LEO)
LEONIS MINORIS (LMI)
On meridian
10 p.m. April 1

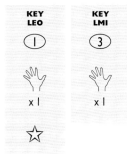

KEY LEO	KEY LMI
①	③

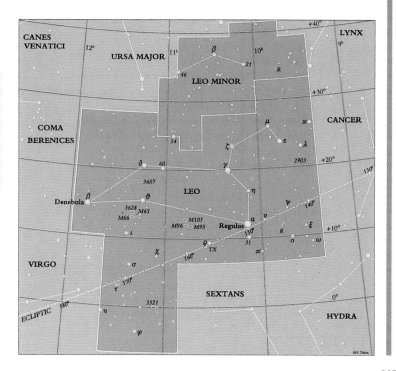

LEPUS (LEE-pus)
The Hare

A faint constellation, Lepus is nevertheless easy to find because it is directly south of Orion. In ancient times it was thought of as Orion's chair. Egyptian observers saw it as the Boat of Osiris, the god who ruled the underworld. The Greeks and Romans gave it the name Lepus. Since Orion particularly liked hunting hares, it was appropriate to place one below his feet in the sky.

Gamma (γ) Leporis Easy to separate in virtually any telescope, this wide double star with contrasting colors has a separation of 96 arc seconds. It is comparatively close to Earth, at a distance of 21 light years, and is part of the Ursa Major stream of stars.

Hind's Crimson Star Likened by some observers to a drop of blood in the sky, R Leporis is the variable that the nineteenth-century British astronomer J. Russell Hind called the Crimson Star. Over a period of 14 months, the star varies in magnitude from a maximum of as much as 5.5 to a minimum of 11.7. Its coloring is at its most striking when the sky is dark and the star is near maximum brightness.

M 79 With the galactic center in Sagittarius half the sky away, this is a surprising place to find a globular cluster. Nevertheless, M 79 is here to enchant you, especially if you have an 8 inch (200 mm) or larger telescope which will begin to resolve the stars around its edges.

A starlit or a moonlit dome distains
All that man is;
All mere complexities,
The fury and the mire of human veins.

BYZANTIUM,
W. B. YEATS (1865–1939),
Irish poet

LEPORIS (LEP)
On meridian
10 p.m. Jan 10

KEY

x 1

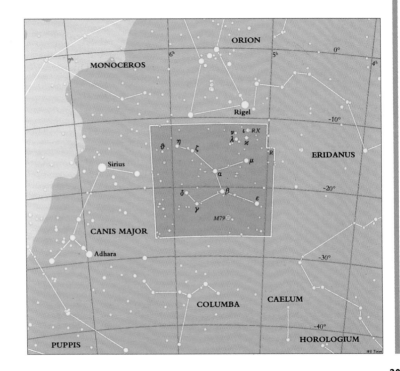

LIBRA (LEE-bra)
The Scales, the Balance

Libra is one of the constellations of the zodiac and was associated with Themis, the Greek goddess of justice, whose attribute was a pair of scales. These stars were also thought of as part of Scorpius: the alpha (α) and beta (β) stars both carry Arabic names, the former being Zuben El Genubi, "southern claw" and the latter Zuben Eschamali, "northern claw." Our understanding is that Libra became a separate constellation at the time of the ancient Romans.

Delta (δ) Librae Similar to Algol, this eclipsing variable star fades by about a magnitude every 2.3 days, from 4.9 to 5.9. The entire cycle is visible to the naked eye.

S Librae A Mira star, S Librae varies from an 8.4 maximum to a 12.0 minimum over a period of just over six months.

FINDING LIBRA
Looking like a high-flying kite, Libra is easy to find by extending a line westward from Antares and its two bright neighbors in Scorpius. The line reaches Alpha (α) and Beta (β) Librae.

LIBRAE (LIB)
On meridian
10 p.m. June 10

KEY

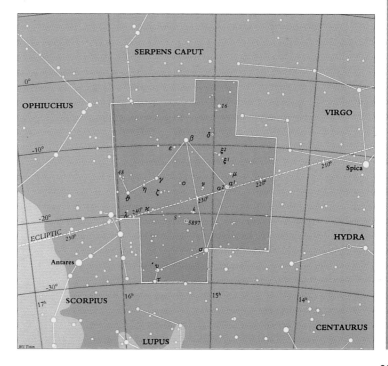

Lupus (LOO-pus)
The Wolf

South of Libra and east of Centaurus, Lupus, the Wolf, is a small constellation with some 2nd magnitude stars. It is almost joined with Centaurus, as if the Centaur is stroking the wolf like a pet. The ancient Greeks and Romans called this group of stars Therion—an unspecified wild animal. Lying within the band of the Milky Way, this constellation is home to a number of open and globular clusters.

RU Lupi (Read this name out loud. After immersing yourself in constellation lore to this point, you possibly are!) RU Lupi is a faint nebular variable, with a maximum of only 9th magnitude. Its irregular variation is characteristic of young stars still involved with nebulosity.

Astronomy compels the soul to look upward and leads us from this world to another.

The Republic, bk. vii,
Plato (c. 428–348 bc),
Greek philosopher

HOW VARIABLE STARS ARE NAMED

Variable stars are named in a strange way. The first variable discovered in a constellation, for example Lupus, is called R Lupi; the second S, then T, until Z. Then the names go RR Lupi, RS to RZ, then SS to SZ, and finally ZZ. In constellations with a great many variables, the list continues with AA to AZ, and on to QZ, but leaving out J. If more than these 334 variables are found, the system continues (more sensibly) with numbers, such as V 1500 Cygni for the 1975 nova in Cygnus.

LUPI (LUP)
On meridian
10 p.m. June 10

KEY

x |

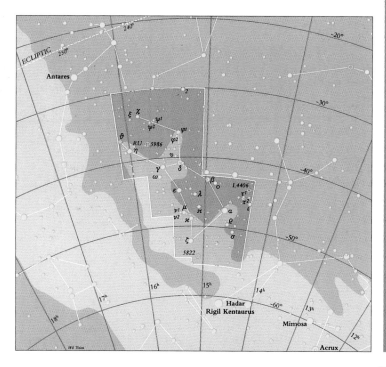

LYNX (LINKS)

The Lynx

With only one 3rd magnitude star, Lynx is one of the hardest constellations to find. Johannes Hevelius charted this figure around 1690, apparently naming it Lynx because you need to have the eyes of a lynx to spot it. The same is true of its deep sky objects.

From the intrinsic evidence of his creation, the Great Architect of the Universe now begins to appear as a pure mathematician.

MYSTERIOUS UNIVERSE,
SIR JAMES JEANS (1877–1946),
English astronomer

The Intergalactic Tramp (NGC 2419)

Lying some 7 degrees north of Castor, the brightest star in Gemini, this is a very faint and distant globular cluster. It is more than 60 degrees from any other globular. At 210,000 light years, it is more distant than the Large Magellanic Cloud and is so far away that it might escape the gravitational pull of our galaxy. It is for this reason that astronomer Harlow Shapley called it the Intergalactic Tramp. Through a 10 inch (250 mm) or larger telescope, NGC 2419 appears as a fuzzy knot of light.

LYNCIS (LYN)
On meridian
10 p.m. Feb 20

KEY

④

x 1.5

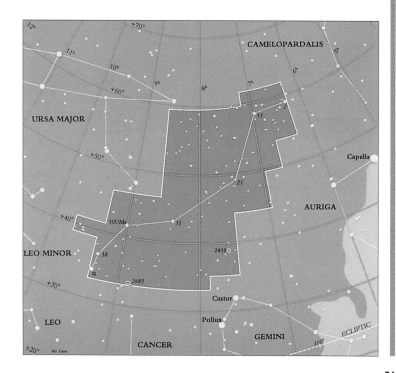

LYRA (LYE-rah)
The Lyre

This beautiful constellation is dominated by Vega, one of the brightest stars in the sky. You can imagine the lyre strings stretched across the parallelogram of four stars that accompany it. The lyre was given by Apollo to his son Orpheus, who played it so exquisitely that wild beasts and the mountains were enchanted.

Stars scribble in our eyes the frosty sagas,
The gleaming cantos of unvanquished space.
HART CRANE (1899–1932),
American poet

Epsilon (ε) Lyrae This is a "double double" star. The slightest optical aid shows two 5th magnitude stars—ϵ^1 and ϵ^2. Both are themselves doubles, with separations under 3 arc seconds. A 4 inch (100 mm) telescope operating at a magnification of 100 or more will split both of them.

Beta (β) Lyrae This eclipsing variable ranges from magnitude 3.3 to 4.4 in 13 days.

The Ring Nebula (M 57) This famous planetary nebula lies midway between Beta (β) and Gamma (γ) Lyrae. Through a 3 inch (75 mm) or larger telescope it appears as a star out of focus at low magnification. Higher power will show its ring shape, about 2 arc minutes across.

LYRAE (LYR)
On meridian
10 p.m. Aug 1

KEY

x 0.5

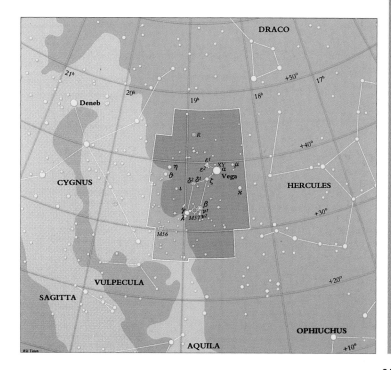

MENSA (MEN-sah)
The Table, the Table Mountain

The only constellation that refers to a specific piece of real estate, Mensa was originally called Mons Mensae by Nicolas-Louis de Lacaille, after Table Mountain, south of Cape Town, South Africa, where he did a good deal of his work. He developed this small constellation from stars between the Large Magellanic Cloud and Octans. The northernmost stars of the constellation, representing the summit of the mountain, are hidden in the Large Magellanic Cloud, in the same way that Table Mountain is often shrouded in clouds.

Alpha (α) Mensae This dwarf star has an apparent magnitude of 5.1. It lies comparatively close to us, its light taking only 28 years to reach Earth.

Beta (β) Mensae Lying very near the edge of the Large Magellanic Cloud, this faint star of magnitude 5.3 lies at a distance of 155 light years away.

MENSAE (MEN)
On meridian
10 p.m. Jan 10

KEY

x 1

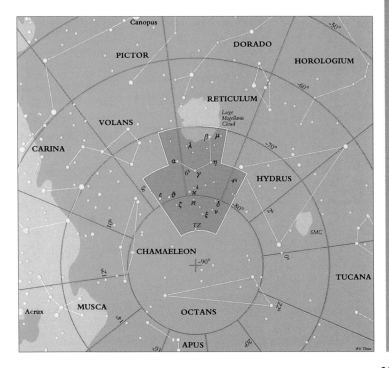

217

MICROSCOPIUM (my-kro-SKO-pee-um)

The Microscope

This small, faint constellation, which lies just south of Capricornus and east of Sagittarius, was created by Nicolas-Louis de Lacaille in about 1750. It commemorates the microscope, the invention of which is credited to the Dutch spectacle-maker Zacharias Janssen, around 1590, and to Galileo, among others.

BR Microscopii This faint Mira variable has a rapid cycle lasting only 4½ months, during which it drops from magnitude 9.2 to 13.4 and climbs back again.

By space the universe encompasses and swallows me up like an atom; by thought I comprehend the world.

PENSÉES,
BLAISE PASCAL (1632–62),
French mathematician and natural philosopher

SKYWATCHING TIP

How good is your telescope? On a night when the atmosphere is steady, you can test its quality by trying to resolve double stars of equal magnitude. Beta (β) Monocerotis, for example, is a triple system. Two stars, of magnitudes 4.7 and 5.2, form a pair separated by 2.8 arc seconds which should be split by a 2.4 inch (60 mm) telescope at 100x magnification. The third star, 7.3 arc seconds away, is even easier to split. Theta 2 (ϑ_2) Microscopii is another triple system but the close pair, separated by 0.5 arc seconds, is a test for a 10 inch (250 mm) telescope in good seeing.

MICROSCOPII (MIC)
On meridian
10 p.m. Sept 1

KEY

x 0.5

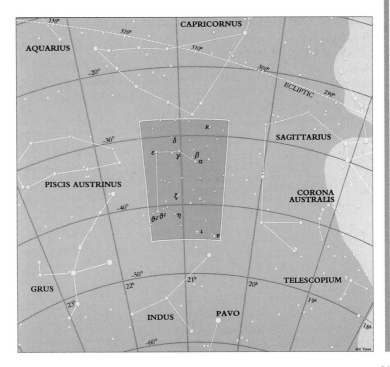

MONOCEROS (moh-NO-ser-us)
The Unicorn

This faint constellation was formed in about 1624 by the German Jakob Bartsch. Monoceros is the Latin form of a Greek word meaning "one-horned," and it seems that the mythical unicorn may have come into existence as a result of a confused description of a rhinoceros.

M 50 This beautiful open cluster, lying slightly more than one-third of the way from Sirius to Procyon, is easy to find. Some of the cluster's stars are arranged in pretty arcs.

The Rosette Nebula (NGC 2237) Through a 10 inch (250 mm) telescope, this ring-shaped nebula, and the open cluster it contains (NGC 2244), offer a scene of delicate beauty. Smaller telescopes and binoculars will reveal the nebula on very clear nights.

The Christmas Tree Cluster (NGC 2264) This open cluster really does resemble a Christmas tree.

This, now, is the judgment of our scientific age—the third reaction of man upon the universe! This universe is not hostile, nor yet is it friendly. It is simply indifferent.

SENSIBLE MAN'S VIEW OF RELIGION
JOHN H. HOLMES (1879–1964),
American clergyman

A WINTER TRIANGLE?
Anxious to create a winter equivalent to the Northern Hemisphere's summer triangle, some observers advocate a winter triangle, bounded by Betelgeuse (in Orion), Sirius (in Canis Major) and Procyon (in Canis Minor). Monoceros, the Unicorn, and the band of the Milky Way fill the space inside this triangle.

MONOCEROTIS (MON)
On meridian
10 p.m. Feb 1

KEY

x 1.5

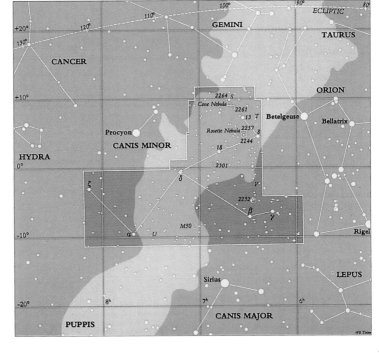

MUSCA (MUSS-kah)
The Fly

Musca is an easy constellation to find, just to the south of the Southern Cross. It was originally described by Johann Bayer in his 1603 star atlas as Apis, the Bee. Later on, Edmond Halley called it Musca Apis, the Fly Bee, then Nicolas-Louis de Lacaille named it Musca Australis, the Southern Fly—to avoid it being confused with the fly on the back of Aries. Now that this northern fly is no longer a constellation, the Southern Fly is known simply as Musca.

On a clear moonless night in midwinter or midsummer, a plume of starlight rises motionless behind the scattering of constellations...The Milky Way is our island universe...
    CHARLES A. WHITNEY (b. 1929),
    American astronomer and writer

Beta (β) Muscae This elegant double star consists of two 4th magnitude stars that revolve around each other in a period that spans several hundred years. The pair is some 520 light years from Earth. The separation of 1.6 arc seconds is very tight, presenting a challenge for a 4 inch (100 mm) telescope.

NGC 4372 This globular cluster is close to Gamma (γ) Muscae and has faint stars spread over 18 arc minutes.

NGC 4833 This is a large, faint globular cluster within 1 degree of Delta (δ) Muscae. A 4 inch (100 mm) or larger telescope is needed to begin to resolve the cluster into individual stars.

MUSCAE (MUS)
On meridian
10 p.m. May 1

KEY

x 0.5

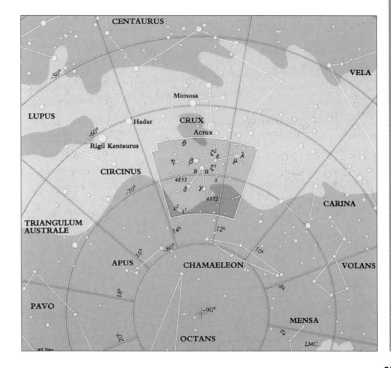

NORMA (NOR-muh)
The Square

East of Centaurus and Lupus is a small constellation called Norma, the Square. When he named this group of stars, Nicolas-Louis de Lacaille decided to call it Norma et Regula, the Level and Square, after a carpenter's tools. Since those days, however, the Regula has been forgotten. The constellation lies alongside Circinus, the Drawing Compass, which he named at the same time. Set in the southern Milky Way, Norma presents good fields for binoculars, with a number of open clusters. For a small constellation, Norma has also been quite lucky with the appearance of novae: there was one in 1893 and another in 1920.

NGC 6067 This is a small open cluster. Large binoculars or a telescope reveal some 100 stars within a stunning field.

NGC 6087 This is another of the constellation's striking open clusters.

NORMAE (NOR)
On meridian
10 p.m. June 10

KEY

x 0.5

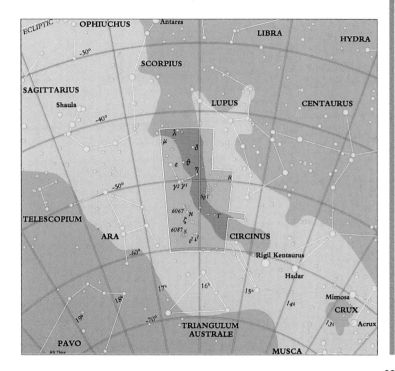

OCTANS (OCK-tanz)
The Octant

To honor John Hadley's invention of the octant in 1730, Nicolas-Louis de Lacaille formed this south polar constellation and called it Octans Hadleianus. The forerunner of the sextant, the octant was an instrument used for measuring the altitude of a celestial body—an essential device for navigators and astronomers.

The stars awaken a certain reverence, because though always present, they are inaccessible...

ESSAY ON NATURE,
RALPH WALDO EMERSON
(1803–82),
American writer

Sigma (σ) Octantis This is the south pole star. At magnitude 5.4, it is barely visible to the naked eye on a dark night, so while it does mark the pole, it is not as convenient a marker star as the north's Polaris. The celestial poles move with time as the axis of the Earth precesses, or wobbles like a top, over some 26,000 years. Sigma (σ) Octantis was at its closest to the pole in about 1870, at just under ½ degree. It is now just over 1 degree. In about another 3,000 years, the pole will begin to move through Carina, and it will pass near Delta (δ) Carinae in about 7,000 years. At 2nd magnitude, this is the brightest south pole star the Earth ever sees.

OCTANTIS (OCT)
On meridian
10 p.m. Sept 10

KEY

x 1

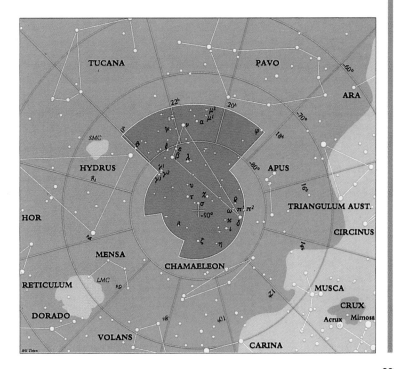

OPHIUCHUS (oh-fee-U-cuss)
The Serpent Bearer

Ophiuchus, entwined with the constellation Serpens, covers a large expanse of sky and contains some of the Milky Way's richest star clouds. Ophiuchus is usually identified with Asclepius, the god of medicine. In one story, Asclepius learned about the healing power of plants from a snake. His skills caused Hades, god of the underworld, to become jealous, and he persuaded Zeus to kill Asclepius. Zeus then placed Asclepius in the sky, along with Serpens, his serpent.

M 9, 10, 12, 14, 19 and 62 These globular clusters provide a range of examples of different concentrations of stars. M 9 and 14 are rich; M 10 and 12 are looser; M 19 is oval; M 62 is somewhat irregular in outline. All are visible in binoculars but require a 6 or 8 inch (150 or 200 mm) telescope to do them justice.

RS Ophiuchi This nova had outbursts in 1898, 1933, 1958, 1967 and 1985. Its minimum is around magnitude 11.8, rising as high as 4.3 during outbursts.

Barnard's Star Discovered by E. E. Barnard in 1916, this 9.5 magnitude red dwarf star has the greatest proper motion (apparent motion across the sky) of any known star. Only 6 light years away, it is the nearest star after the Alpha (α) Centauri system.

KEPLER'S STAR
On October 9, 1604, Ophiuchus hosted our galaxy's most recent supernova. Known as Kepler's Star, it outshone Jupiter for several weeks.

Behold, directly overhead, a certain strange star was suddenly seen . . . Amazed, and as if astonished and stupified, I stood still.

TYCHO BRAHE (1546–1601),
Danish astronomer

OPHIUCHI (OPH)
On meridian
10 p.m. July 10

KEY

②

x 2

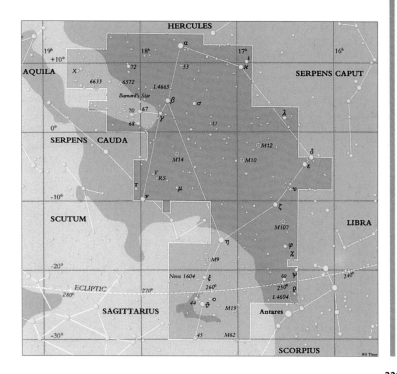

ORION (oh-RYE-un)
The Hunter

Orion is a treasure, with Rigel, Betelgeuse and its three belt stars in a row lighting up the sky from December to April. The stars are arranged so that it is easy to see the figure of a hunter, complete with lion's skin in one hand and raised club in the other.

👁 Betelgeuse (Alpha [α] Orionis)

Betelgeuse (pronounced BET-el-jooze but sometimes corrupted to BEETLE-juice) is fabulous. (Its name comes from the Arabic for "house of the twins," apparently because of the adjacent constellation of Gemini.) A variable star, it varies in magnitude from 0.3 to 1.2 over a period of almost seven years. However, the semi-regular nature of the variation means that it is often possible to detect changes over just a few weeks.

👁 Rigel (Beta [β] Orionis)

The name Rigel is derived from the Arabic for "foot." This mighty supergiant, which is about 1,400 light years away, is more than 50,000 times as luminous as the Sun.

👁 The Great Nebula (M 42)

This star nursery, one of the marvels of the night sky, is also known as the Orion Nebula. Plainly visible to the naked eye under a dark sky, it can be clearly seen through binoculars even in the city. The swirls of nebulosity spread out from its core of four stars called the Trapezium, which power the nebula. Photographs usually "burn-out" the inner region of the nebula and obscure the Trapezium stars. In 1880, using M 42 as his subject, Henry Draper was the first person to successfully photograph a nebula.

🔭 M 43

This is a small patch of nebulosity just north of the main body of the Great Nebula. In fact, the M 42 complex is simply the brightest part of a gas cloud covering the constellation of Orion at a distance of some 1,500 light years.

🔭 The Horsehead Nebula (IC 434)

Also known as Barnard 33, this dark nebula is projected against a background of diffuse nebulosity, alongside the bright belt star Zeta (ζ) Orionis. It can be quite difficult to see, usually requiring a dark sky and at least an 8 inch (200 mm) telescope.

🔭 NGC 2169

This is a small, bright open cluster made up of about 30 stars.

ORIONIS (ORI)
On meridian
10 p.m. Jan 10

KEY

x 1.5

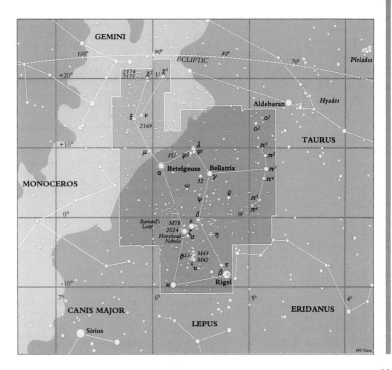

PAVO (PAH-voh)
The Peacock

Pavo, the Peacock, lies not far from the south celestial pole, south of Sagittarius and Corona Australis. It is a modern constellation, devised by Johann Bayer and published in his star atlas of 1603, but he may have been thinking of the mythical peacock that was sacred to Hera, the goddess of women and marriage in Greek mythology.

👁 The Peacock Star (Alpha (α) Pavonis)
This star is 150 light years away. It is a binary system whose members orbit each other in less than two weeks, but the pair is too close to separate telescopically.

NGC 6752
A spectacular globular cluster at a relatively close distance of 17,000 light years, this huge family of stars is the third largest globular cluster (in apparent size) after Omega (φ) Centauri and 47 Tucanae.

NGC 6744
This faint but beautiful galaxy is one of the largest known barred spirals. Smaller telescopes reveal only the nuclear regions, a 10 inch (250 mm) telescope being necessary to reveal anything more.

In winter the stars seem to have rekindled their fires, the moon achieves a fuller triumph, and the heavens wear a look of more exalted simplicity.

JOHN BURROUGHS (1837–1921),
American naturalist

PAVONIS (PAVI)
On meridian
10 p.m. Aug 10

KEY

x 1

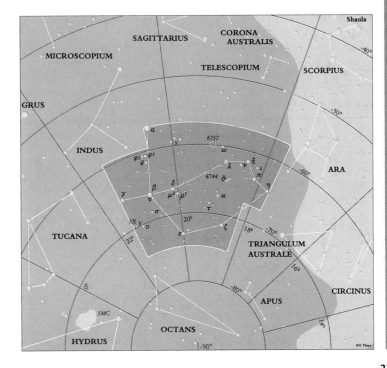

PEGASUS (PEG-a-sus)
The Winged Horse

Although this constellation has no really bright stars, it is easy to spot because its three brightest stars, with Alpha (α) Andromedae, form the Great Square of Pegasus. The winged horse has been found on ancient tablets from the Euphrates, and on Greek coins of the fourth century BC. According to Greek legend, when Perseus decapitated the Gorgon Medusa, Pegasus sprang up from her blood. When Pegasus was brought to Mount Helicon, one kick of his hoof caused the spring of Hippocrene to flow—source of inspiration for poets.

 M 15 One of the best of the northern sky globular clusters, M 15 is 34,000 light years away. Although it is visible through binoculars as a nebulous patch, in a telescope it is a real showpiece.

NGC 7331 This spiral galaxy is the brightest one in Pegasus, but is still only 9th magnitude.

Stephan's Quintet This very faint group of galaxies lies ½ degree south of NGC 7331. Even though faint streamers of material seem to connect the largest of the galaxies to the others, detailed study indicates that this galaxy is probably closer to us than they are. These galaxies are not really targets for the beginner, as they need at least a 10 inch (250 mm) telescope to be seen clearly.

PEGASI (PEG)
On meridian
10 p.m. Oct 1

KEY

x 2

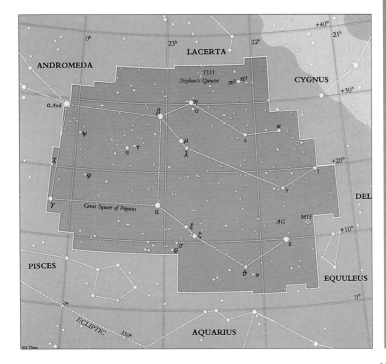

PERSEUS (PURR-see-us)
The Hero

A pretty constellation that straddles the Milky Way, Perseus is in the northern skies from July to March. Its stars arc from Capella, in Auriga, to Cassiopeia. The son of the chief god Zeus and the mortal Danaë, Perseus's most famous exploit was to kill the Gorgon Medusa—one of three sisters that were so terrifyingly ugly that one glance of them would turn the viewer to stone. Using the shield of Athene as a mirror, he severed Medusa's head, and the winged horse Pegasus sprang from her blood.

Algol The star that winks, this is the most famous of the eclipsing variables. Every 2 days, 20 hours and 48 minutes, it begins to drop in brightness from magnitude 2.1 to 3.4 in an eclipse lasting 10 hours.

M 34 This bright open cluster sits in the middle of a rich field of stars. It is an interesting view through either binoculars or a telescope.

Double Cluster (NGC 869 and 884) Two of the finest examples of open clusters in the sky, NGC 869 and 884 (h Persei and Chi [χ] Persei respectively) are magnificent through binoculars or the low-power field of a small telescope.

Perseids One of the best meteor showers, these meteors, which come from periodic comet Swift-Tuttle, peak on August 11 and 12.

SKYWATCHING TIP

You do not need a telescope or binoculars to observe a meteor shower such as the Perseids. The naked eye, in fact, makes the best instrument because of the large amount of sky it can take in at any one time. However, binoculars can help you locate fainter meteors. Meteor watching often involves long sessions, so it is important to be comfortable. A deck chair is an easy way to keep your head inclined at the right angle. Position yourself so that you face the shower's radiant—the point in the sky where the shower appears to come from—but do not ignore other parts of the sky.

PERSEI (PER)
On meridian
10 p.m. Dec 10

KEY

x 1.5

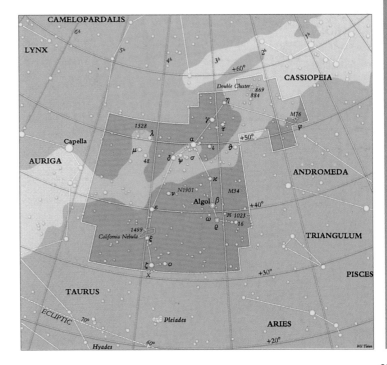

PHOENIX (FEE-nicks)
The Phoenix

A fabulous symbol of rebirth, in mythology the Phoenix was a bird of great beauty that lived for 500 years. It would then build a nest of twigs and fragrant leaves which would be lit by the noontime rays of the Sun. The Phoenix would be consumed in the fire, but a small worm would wriggle out from the ashes, bask in the Sun and quickly evolve into a brand new Phoenix. The constellation first appeared in Bayer's *Uranometria* of 1603.

Look at the stars! Look, look up at the skies!

O look at all the fire-folk sitting in the air!

The bright boroughs, the circle-citadels there!

THE STARLIGHT NIGHT,
GERARD MANLEY HOIPKINS
(1844–89),
English poet

SX Phoenicis The best example of a "dwarf Cepheid" variable, this star changes from magnitude 7.1 to 7.5 and back again in only 79 minutes and 10 seconds! Cepheid periods are very exact. In this case, however, the range varies, with some maxima as bright as 6.7. The variation probably occurs because the star has two different oscillations occurring at once. Such a small range in brightness can be difficult to monitor, requiring careful comparison with neighboring stars.

PHOENICIS (PHE)
On meridian
10 p.m. Nov 1

KEY

x 1.5

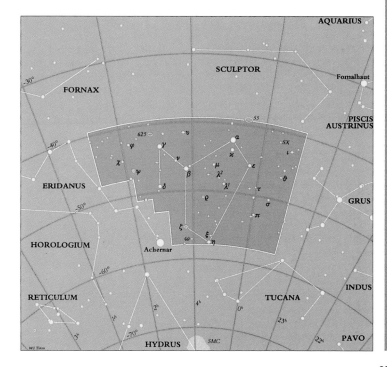

PICTOR (PIK-tor)
The Painter's Easel

This southern constellation was originally named Equuleus Pictoris, the Painter's Easel, by Nicolas-Louis de Lacaille. Nowadays its shortened name refers solely to the painter. It is a dull group of stars lying south of Columba and alongside the brilliant star Canopus.

👁 Beta (β) Pictoris This 4th magnitude star is host to a disk of dust and ices which could be a planetary system in formation. The surrounding nebula is only visible using special techniques on large telescopes.

Kapteyn's Star Only 12.7 light years away, this star was discovered by the Dutch astronomer Jacobus Kapteyn in 1897. It moves quickly among distant background stars, crossing 8.7 arc seconds of sky per year—the width of the Moon every two centuries. At magnitude 8.8, the star is visible through binoculars and small telescopes.

PICTORIS (PIC)
On meridian
10 p.m. Jan 10

KEY

③

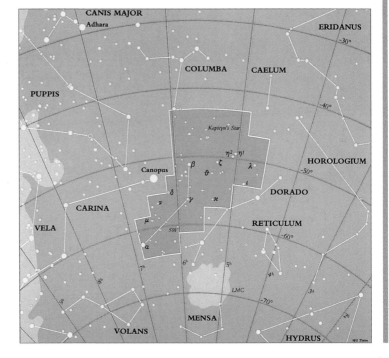

PISCES (PIE-seez)
The Fish

For thousands of years, this faint zodiacal constellation has been seen either as one fish or two. In Greco-Roman mythology, Aphrodite and her son Heros were at one time being pursued by the monster Typhon. To escape him, they turned themselves into fish and swam away, having tied their tails together to make sure that they would not be parted. The ring of stars in the western fish, which is beneath Pegasus, is called the Circlet. The eastern fish is beneath Andromeda.

Zeta (ζ) Piscium A beautiful double star of magnitudes 5.6 and 6.5, separated by 24 arc seconds.

M 74 This is a large spiral galaxy, seen face-on, close to Eta (η) Piscium. While it is the brightest Pisces galaxy, it is still rather faint and requires a dark sky and an 8 inch (200 mm) telescope or larger to be seen.

Van Maanen's Star This is a rare example of a white dwarf star that, at magnitude 12.2, can actually be identified in an 8 inch (200 mm) telescope.

THE MESSIER MARATHON

The eighteenth-century French astronomer Charles Messier listed all the objects he encountered—mainly clusters, nebulae and distant galaxies—while looking for comets. In the Northern Hemisphere, a window of opportunity opens around the time of the vernal equinox, every March, when all 110 Messier objects can be observed on one night. To complete a Messier Marathon, begin with the spiral galaxy M 74 in Pisces, in the western sky at dusk. During the night, sweep from one Messier object to the next, finishing with M 30 in Capricornus, shortly before the onset of morning twilight.

PISCIUM (PSC)
On meridian
10 p.m. Nov 1

KEY

x 2.5

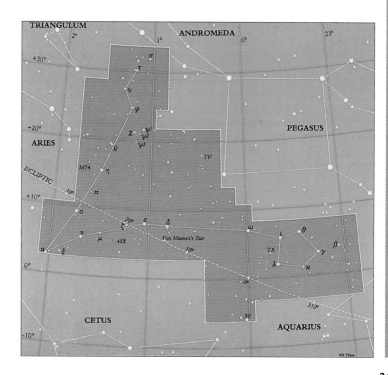

PISCIS AUSTRINUS (PIE-sis OSS-trih-nuss)
The Southern Fish

Lying to the south of Aquarius and Capricornus, Piscis Austrinus, the Southern Fish, is relatively easy to spot because of its lone, bright star, Fomalhaut, which is often referred to as the Solitary One. For the Persians, 5,000 years ago, this was a Royal Star that had the privilege of being one of the guardians of heaven. Many early charts of the heavens show the Southern Fish drinking water that is being poured from Aquarius's jar.

◉ Fomalhaut At magnitude 1.2, this star is 22 light years away—close by stellar standards. It is about twice as large as our Sun and has 14 times its luminosity. Some 2 degrees of arc southward is a magnitude 6.5 dwarf star that seems to be sharing Fomalhaut's motion through space. They are so far apart that it is hard to call them a binary system. Maybe these two stars are all that is left of a cluster that dissipated long ago.

To persons standing alone on a hill during a clear midnight such as this, the roll of the world eastward is almost a palpable movement.

FAR FROM THE MADDING CROWD
THOMAS HARDY (1840–1928),
English poet and novelist

PISCIS AUSTRINI (PSA)
On meridian
10 p.m. Oct 1

KEY

x 1

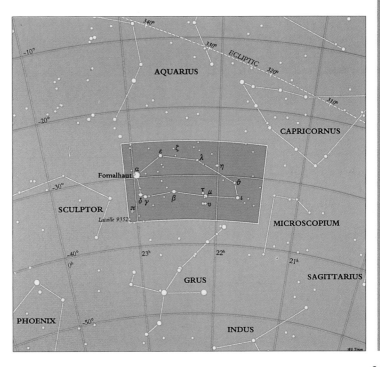

245

PUPPIS (PUP-iss) & PYXIS (PIK-sis)
The Stern & the Compass

Found just south of Canis Major, Puppis is the stern of the ship *Argo*. It is the northernmost of the constellations that formed the ship. With the Milky Way running along it, Puppis provides a feast of open clusters for binoculars or telescopes. Right alongside is the smaller and fainter constellation of Pyxis, which used to be Malus, *Argo*'s mast, before Lacaille made a ship's compass out of it.

Zeta (ζ) Puppis This blue supergiant sun is one of our galaxy's largest. About 2,000 light years away, it shines at 2nd magnitude.

L² Puppis One of the brightest of the red variable stars, L² Puppis varies from magnitude 2.6 to 6.2 over a period of five months.

M 46 A beautiful open star cluster, through small telescopes M 46 is a circular cloud of faint stars the apparent diameter of the Moon. A faint planetary nebula, NGC 2438, appears to be a part of the cluster but is not a true member. It is 11th magnitude and 1 arc minute across, needing an 8 inch (200 mm) or larger telescope to be seen well.

T Pyxidis A recurrent nova with a faint 16th magnitude minimum, T Pyxidis sometimes reaches 7th magnitude during its outbursts, which occur at intervals of 12 to 25 years.

SKYWATCHING TIP

If you think you have seen a nova or supernova, note its position and estimate its magnitude. Always double-check your observation, and have it confirmed by another competent observer, an observing organization, or an observatory. Give them the date and time of your observation, the instrument used, the position and brightness of the object, and the observing conditions.

PUPPIS (PUP)
PYXIDIS (PYX)
On meridian
10 p.m. Feb 20

KEY PUP	KEY PYX
③	④

x 1.5   x 0.5

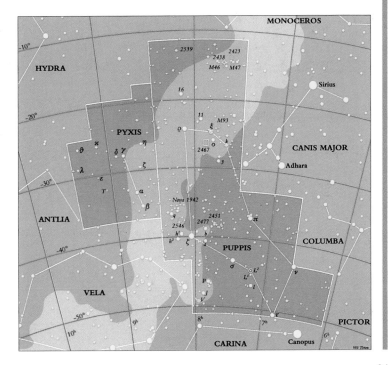

Wil Tirion

Sky and Constellation Charts

247

RETICULUM (reh-TIK-u-lum)
The Reticle

A small constellation of faint stars halfway between the bright stars Achernar and Canopus, Reticulum was first set up as Rhombus by Isaak Habrecht of Strasburg. De Lacaille changed its name to Reticulum to honor the reticle—the grid of fine lines in a telescope eyepiece that aids with centering and focussing. It is occasionally also known as the Net.

R Reticuli This Mira star is quite red, and at maximum light it shines at about magnitude 7 Over a period of nine months, it drops to magnitude 13, then returns to maximum brightness.

SKYWATCHING TIP

No matter what instrument you use, dark, clear skies are essential for deep-sky astronomy. Some galaxies are only slightly brighter than the normal background skyglow, so you need all the contrast you can muster from your instrument and your eyes. Before you begin observing, give your eyes 15 or 20 minutes to become dark-adapted. Then use a red-filtered flashlight to study star charts without ruining your night vision.

Though the day still lingers, the rose-scattering fire of the evening star already scintillates through the azure sky.

WILLEM KLOOS (1859–1938),
Dutch poet and essayist

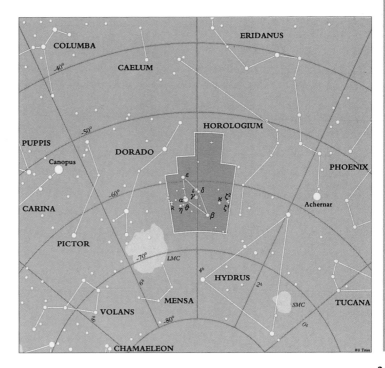

SAGITTA (sa-JIT-ah)
The Arrow

Although only a small constellation, Sagitta is easy to find halfway between Altair in Aquila, and Albireo (Beta [β] Cygni). Its shape really is true to its name—the ancient Hebrews, Persians, Arabs, Greeks and Romans all saw this group of stars as an arrow. It has been thought of variously as the arrow that Apollo used to kill the Cyclops; one of the arrows shot by Hercules at the Stymphalian Birds; and as Cupid's dart.

Old men and comets have been reverenced for the same reason, their long beards and pretence to foretell events.

JONATHAN SWIFT (1667–1745),
Anglo-Irish writer

U Sagittae Every 3.4 days, this eclipsing binary drops from magnitude 6.5 to a minimum of 9.3.

V Sagittae Although this star is faint, varying erratically from magnitude 8.6 to magnitude 13.9, it is interesting because of the way it alters a little almost every night. It might have been a nova a long time ago.

M 71 A little south of the midpoint of a line joining Delta (δ) and Gamma (γ) Sagittae, M 71 is a fertile cluster of faint stars. It is now generally regarded as a poor, uncondensed globular cluster, rather than as a rich open cluster.

SAGITTAE (SGE)
On meridian
10 p.m. Aug 20

KEY

x |

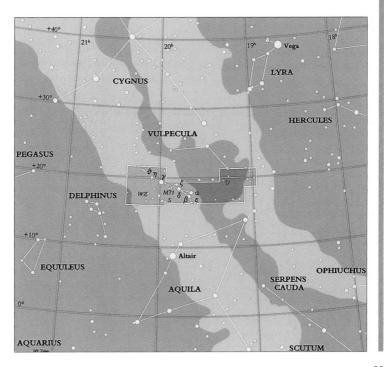

SAGITTARIUS (sadge-ih-TAIR-ee-us)
The Archer

Sagittarius is generally thought to be a centaur—half man and half horse—and is usually considered to be Chiron, who is also identified with the constellation Centaurus. The constellation is located on the Milky Way in the direction of the center of the galaxy. Here the band of the Milky Way is at its broadest, although cut by dark bands of dust. It is a treasure trove of galactic and globular clusters, plus bright and dark nebulae.

M 22 The Great Sagittarius star cluster is a very large globular—the best of the constellation's many globulars. At magnitude 6.5, it is an easy object to see in binoculars, but a telescope really brings out the cluster's beauty. Only 10,000 light years away, it is one of the closest globulars, and with an 8 inch (200 mm) telescope you should be able to resolve it into seemingly countless stars.

M 23 Just one of many galactic clusters in Sagittarius, M 23 presents over 100 stars in an area the size of the Moon. It is a striking sight in binoculars or in a telescope at low magnification.

The Lagoon Nebula (M 8)
This spectacular diffuse nebula envelops the cluster of stars called NGC 6530. On a dark night the nebula is visible to the naked eye just to the north of the richest part of the Sagittarius Milky Way. In photographs, the extensive nebula is marked by several tiny dark splotches. Dutch astronomer Bart Bok identified these as globules in which new stars are being formed.

The Trifid Nebula (M 20)
Found only 1½ degrees to the northwest of the Lagoon Nebula, the Trifid Nebula is likely to be part of the same complex of nebulosity. It is known as the Trifid because three lanes of dark clouds divide the nebula in the most beautiful way. You should be able to detect these dark lanes with a 6 inch (150 mm) telescope under a good sky.

The Omega Nebula (M 17) Also called the Swan, the Horseshoe or the Check-mark, this nebula can be seen quite clearly in binoculars and is a stunning sight through a large telescope.

KEY

x 1.5

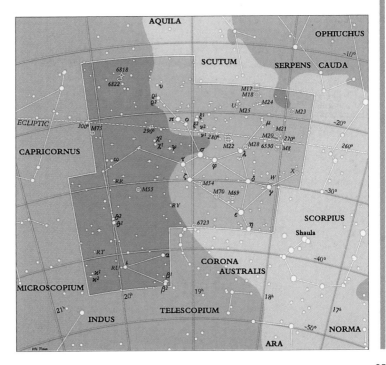

SCORPIUS (SKOR-pee-us)
The Scorpion

A beautiful constellation of the zodiac, filled with bright stars and rich star fields of the Milky Way, Scorpius really looks like a scorpion, complete with head and stinger. Near the northern end is a line of three bright stars, with red Antares (Greek for "rival of Mars") at its center.

Antares The Romans called this star Cor Scorpionis, meaning "heart of the scorpion," a title the French also use—Le Coeur de Scorpion. About 520 light years away, Antares is a red supergiant 600 million miles (1,000 million km) across and is 9,000 times more luminous than the Sun. However, with a mass only 10 or 15 times that of the Sun, it is not very dense.

Beta (β) Scorpii This is a double star whose 2.6 and 4.9 magnitude components are 13.7 arc seconds apart, making resolution possible in a 2 inch (50 mm) telescope.

M 4 "There are several M 4s," said the astronomer Walter Scott Houston. What he meant was that this strange globular cluster has a different appearance with each instrument you use. Binoculars show a fuzzy patch of light; a small telescope shows a large patch of mottled haze; and 4 or 6 inch (100 to 150 mm) instruments begin to show the individual stars. This is one of the best globulars for viewing in small telescopes.

The Butterfly Cluster (M 6) The stars of this large, bright open cluster really resemble a butterfly when viewed at high power.

M 7 This large, bright open cluster, lying to the southeast of M 6, needs to be seen through the large field of view of binoculars to be fully appreciated.

NGC 6231 Half a degree north of Zeta (ζ) Scorpii, this bright open cluster lies in a rich region of the Milky Way. It is best surveyed in binoculars or at very low power in a telescope.

M 80 This bright globular cluster can be seen in binoculars but needs a 10 inch (250 mm) telescope to resolve its stars.

Scorpius X-1 This is a close binary star in which one star expels gas onto a dense neighbor which could be either a white dwarf, a neutron star or a black hole. It is a bright source of X-rays, but appears visually as a 13th magnitude star.

SCORPII (SCO)
On meridian
10 p.m. July 1

KEY

x 1.5

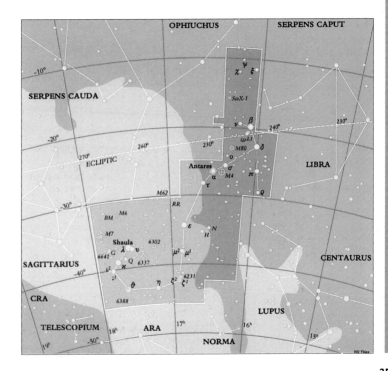

SCULPTOR (SKULP-tor)
The Sculptor

This constellation, originally named L'Atelier du Sculpteur (the Sculptor's Workshop) by Nicolas-Louis de Lacaille in the eighteenth century, lies to the south of Aquarius and Cetus. Since that time, its name has been shortened to Sculptor. Its most significant feature is a small cluster of nearby spiral galaxies.

NGC 253 For a small telescope user, this magnitude 7 galaxy is one of the most satisfying, especially for observers in the Southern Hemisphere. It is very large and is viewed almost edge-on. It was discovered by Caroline Herschel one night in 1783, while she was searching for comets. It appears as a thick streak in binoculars and begins to show the texture evident in photographs when larger instruments are used. The galaxy is 10 million light years distant.

NGC 55 This is another very fine edge-on galaxy, similar to NGC 253. It is distinctly brighter at one end than the other when seen through an 8 inch (200 mm) telescope. Both NGC 55 and 253 are members of the Sculptor Group, an arrangement of several galaxies that might be our Local Group's nearest neighbor in the cosmos.

...the sky
Spreads like an ocean hung on high,
Bespangled with those isles of light...

SIEGE OF CORINTH,
LORD BYRON (1788–1824),
English poet

KEY

x 1.5

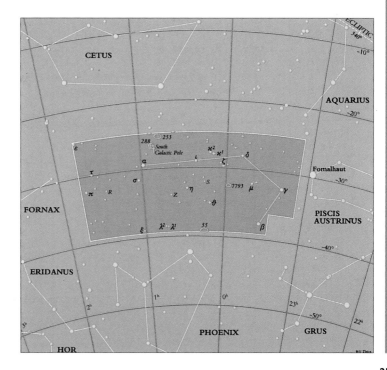

SCUTUM (SKU-tum)
The Shield

Although Scutum is not a large constellation and has no bright stars, it is not difficult to find in a dark sky because it is the home of one of the Milky Way's most dramatic clouds of stars. Johannes Hevelius created the constellation at the end of the seventeenth century, giving it the name Scutum Sobiescianum (Sobieski's Shield) in honor of King John Sobieski of Poland, after he had successfully fought off a Turkish invasion in 1683.

 R Scuti This is a semi-regular RV Tauri-type variable star. It changes from magnitude 5.7 to 8.4 and back over a period of about five months.

The Wild Duck Cluster (M 11) This spectacular open cluster is clearly visible in binoculars, rewarding in a small telescope, and stunning in an 8 inch (200 mm) one. One of the most compact of all the open clusters, the presence of a bright star in the foreground adds to its beauty.

The sun and stars that float in the open air,
The apple-shaped earth and we upon it,
surely the drift of them is something grand.

WALT WHITMAN (1819–92),
American poet

SCUTI (SCT)
On meridian
10 p.m. Aug 1

KEY

x 0.5

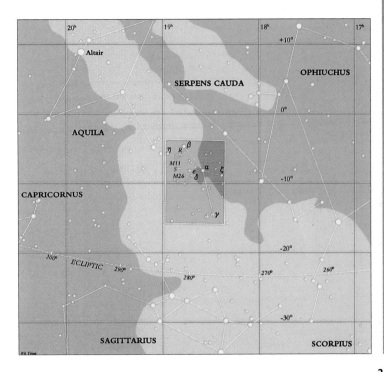

Sky and Constellation Charts

SERPENS (SIR-penz)
The Serpent

This is the only constellation that is divided into two parts. The head (Serpens Caput) and the tail (Serpens Cauda) are separated by the constellation of Ophiuchus, the Serpent Bearer. At one time both the Serpent and the Serpent Bearer formed a single constellation. Serpens was familiar to the ancient Hebrews, Arabs, Greeks and Romans.

R Serpentis A Mira star almost midway between Beta (β) and Gamma (γ) Serpentis, this variable has a bright maximum of 6.9. It fades to about 13.4, although it sometimes can become fainter. Its period is about one year.

M 5 This very striking globular cluster is about 26,000 light years away.

The Eagle Nebula (M 16) Through an 8 inch (200 mm) or larger telescope on a dark night, this combination of nebula and star cluster is quite stunning. But you can still enjoy the sight of the cluster in smaller telescopes.

SKYWATCHING TIP
Any telescope in the 4 to 6 inch (100 to 150 mm) range can show prominent nebulae in dark, clear skies. However, the light grasp of larger scopes is a definite advantage when looking for less substantial nebulae. Once you have located a nebula, boost your magnification for greater contrast until the image begins to degrade.

SERPENTIS (SER)
On meridian
10 p.m. Jun 20 (S. Caput)
July 20 (S. Cauda)

KEY

x 3

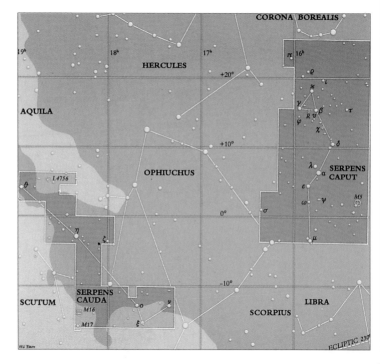

Sky and Constellation Charts

SEXTANS (SEX-tanz)
The Sextant

Sextans Uraniae, now known simply as Sextans, was the creation of Johannes Hevelius. He chose this name for the constellation to commemorate the loss of the sextant he once used to measure the positions of the stars. Along with all his other astronomical instruments, the sextant was destroyed in a fire that took place in September 1679. "Vulcan overcame Urania," Hevelius remarked sadly, commenting on the fire god having defeated astronomy's muse.

HEAVENLY MINISTER

Placed between Leo and Hydra, Sextans' brightest star is barely visible to the unaided eye at magnitude 4.5. Despite this, in ancient times the Chinese chose one of the faintest stars in Sextans to represent Tien Seang, the Minister of State in Heaven.

The Spindle Galaxy

(NGC 3115) Because we see this 10th magnitude galaxy almost edge-on, it appears to be shaped like a lens. Unlike many faint galaxies, the Spindle Galaxy gives quite satisfying views at high power. It seems to be somewhere between an elliptical and a spiral.

SEXTANTIS (SEX)
On meridian
10 p.m. Mar 20

KEY

x 1

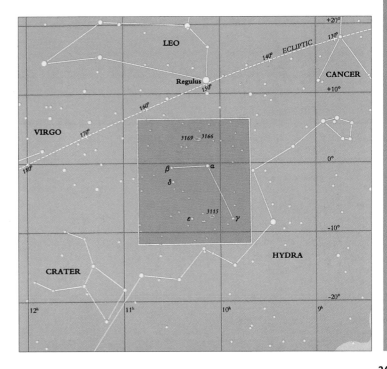

TAURUS (TORR-us)
The Bull

The Greeks saw this constellation as Zeus disguised as a bull. The story goes that Zeus fell in love with Europa, daughter of Agenor, king of Phoenicia. One day, playing at the water's edge, Europa's gaze was caught by a majestic white bull—Zeus in animal form—grazing peacefully among her father's herd. The bull knelt before her as she approached it, so she climbed on its back, wreathing flowers around its horns. Springing to its feet, the bull then took off into the sea and swam to Crete, where Zeus made Europa his mistress.

The Pleiades (M 45) Also known as the Seven Sisters, this is the most famous open star cluster in the sky and forms the bull's shoulder. Greek legend tells that the sisters called for help from Zeus when they were being pursued by Orion. Zeus turned them into doves and placed them in the sky. Alcyone (Eta [η] Tauri) is the most dazzling sister. She is accompanied by: Maia (20 Tauri); Asterope I and II (the double star 21 Tauri); Taygeta (19 Tauri); Celaeno (16 Tauri); and Electra (17 Tauri).

Finally, there is Merope (23 Tauri), a star surrounded by a beautiful cloud of cosmic grains producing a blue reflection nebula. Atlas (or Pater Atlas, 27 Tauri) and Pleione (Mater Pleione, 28 Tauri) represent the girls' father and mother. On a reasonably dark night, you should be able to see at least six of the stars in the Pleiades with the naked eye; under good conditions, you might be able to see as many as nine. Containing more than 500 stars, the Pleiades is about 400 light years away and covers an area four times the size of the full Moon. It is best seen with binoculars.

The Hyades Like the Pleiades, this is also an open cluster, but it is so close to us (only 150 light years away) that even when viewed with the naked eye the stars appear to be spread out. The stars of the Hyades form the bull's head.

Aldebaran (Alpha (α) Tauri) This is an orange giant and is the brightest star in Taurus. Its name means "the follower" (of the Pleiades) in Arabic. Only 60 light years away, it marks the bull's eye.

The Crab Nebula (M 1) This nebula marks the site of the supernova seen in 1054. It is clearly visible through a 4 inch (100 mm) telescope on a dark night as an oval glow, 5 arc minutes across. But this only hints at the complex structure that can be seen in high-magnification photographs.

TAURI (TAU)
On meridian
10 p.m. Jan I

KEY

x 2

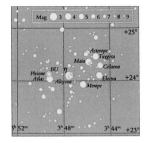

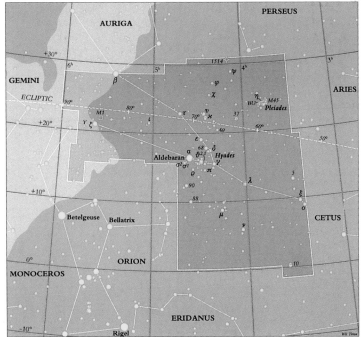

265

TELESCOPIUM (tel-eh-SKO-pee-um)
The Telescope

Originally bearing the name Tubus Telescopium, this constellation was created by Nicolas-Louis de Lacaille during the eighteenth century to honor the invention of the telescope. It was the only large telescope in space until the launch of the Hubble Space Telescope in 1990. Telescopium is surrounded by Sagittarius, Ophiuchus, Corona Australis and Scorpius, and de Lacaille "borrowed" a number of stars from these constellations in order to create it.

Sweet the coming on
Of grateful evening mild, then
*   silent night*
With this her solemn bird and
*   this fair moon.*
And these the gems of heaven,
*   her starry train.*

PARADISE LOST,
JOHN MILTON (1608–74),
English poet

RR Telescopii Although this star is normally too faint for small telescopes, it is one of the most interesting novae on record. Before 1944, this star varied over about 13 months between 12.5 and 15th magnitude, but in that year it began a rise to magnitude 6.5 that took some five years. As the nova declined in the following years, it still displayed its original 13-month period. It is thought that the star may be a binary system, in which a large red star is responsible for the minor variations that take place, and a smaller, hotter star puts on the nova part of the performance.

TELESCOPII (TEL)
On meridian
10 p.m. Aug 10

KEY

x 2

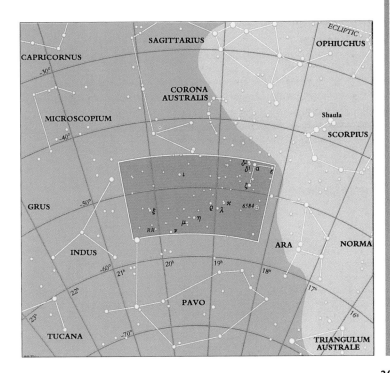

TRIANGULUM (tri-ANG-gyu-lum)
The Triangle

Triangulum is a small, faint constellation extending just south of Andromeda, near Beta (β) and Gamma (γ) Andromedae. Despite its lack of distinction, the group of stars was known to the ancients, and because of its similarity to the Greek letter delta (Δ) it was sometimes called Delta or Deltotum. It has been associated with the delta of the River Nile and has also been connected with the island of Sicily, which is shaped like a triangle. The ancient Hebrews gave it the name of a triangular musical instrument.

👁 The Pinwheel Galaxy
(M 33) This galaxy is one of the brightest and biggest members of our Local Group and we have a front row view because it appears face-on. The galaxy is listed at magnitude 5.5 but its light is spread out over such a large area that it is notoriously difficult to see. Although it can be seen by the naked eye on very clear nights, you need a dark sky and binoculars to see a fuzzy glow larger than the apparent diameter of the Moon. A telescope with a wide field of view will also show the galaxy, but one with a narrow field will show nothing at all.

SKYWATCHING TIP

One of the joys of stargazing is to sit outside under a very dark sky and watch the broad swath of the faintly glowing Milky Way slowly pass overhead. Take the time to study the Milky Way without binoculars or telescope. Look for the delicate interweaving of the bands of obscuring dust clouds and the hazy star fields. After observing the Milky Way with the naked eye, scan the band with binoculars, stopping to study the rich fields and coal-black dark nebulae throughout it. The plane of the Milky Way galaxy is made up of vast clouds of glowing gas and stars cut by dark lanes of dust. Billions of faint stars, many that are far too faint for our eyes to separate into individual points of light, merge like a mottled, grayish fog.

TRIANGULI (TRI)
On meridian
10 p.m. Nov 20

KEY

③

×｜

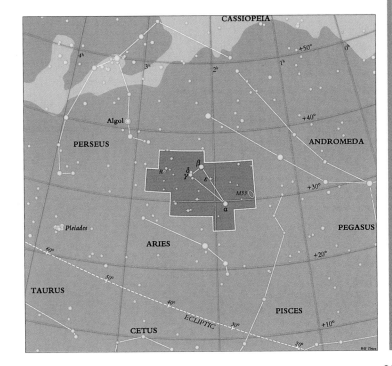

TRIANGULUM AUSTRALE (tri-ANG-gyu-lum os-TRAH-lee)

The Southern Triangle

A simple three-sided figure deep in the southern sky, Triangulum Australe first appeared in Johann Bayer's great atlas *Uranometria* in 1603. It lies just south of Norma, the Level, and east of Circinus, the Drawing Compass—tools used by woodworkers and navigators on early expeditions to the Southern Hemisphere.

R Trianguli Australis One of several Cepheids in the constellation, this interesting variable alters by about a magnitude—from 6.0 to 6.8. Because it is a Cepheid variable, we know its period precisely, which is 3.389 days. For Cepheids with this rapid variation, magnitude estimates at least once a night are worthwhile.

S Trianguli Australis Another bright Cepheid variable, S varies from magnitude 6.1 to 6.7 and back over a period of 6.323 days.

NGC 6025 This is a small open cluster of about 30 stars of 9th magnitude, with fainter background stars.

SKYWATCHING TIP

Should you become serious about observing, you may find it useful to build an observatory in your backyard—somewhere you can keep your telescope set up and ready for use. A number of companies sell prefabricated structures, ranging from modified tents to simple metal or fiberglass buildings. You can even buy a ready-made dome. If you know your way around a hammer and screwdriver, you could make a simple observatory out of a garden shed. With such a design, when the roof is slid back you will be able to swing your telescope around to see the full sky rather than just a section of it.

TRIANGULI AUSTRALIS
(TRA)
On meridian
10 p.m. June 20

KEY

x |

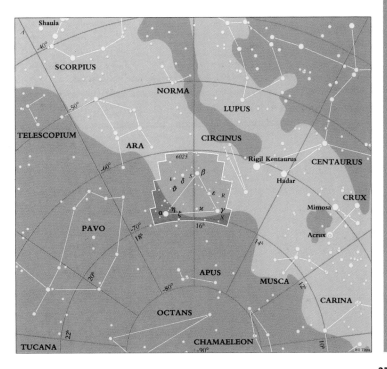

TUCANA (too-KAN-ah)
The Toucan

Johann Bayer first published this constellation as Toucan, which in time became Tucana, the Latin form. Toucans are larger members of the genus *Ramphastos*—brightly colored, large-billed birds related to woodpeckers that are found in tropical America. From the earliest drawings, Tucana sat on the Small Magellanic Cloud, one of the two closest galaxies to the Milky Way, tending it like an egg.

Scientist alone is the true poet
he gives us the moon
he promises the stars
he'll make us a new universe
if it comes to that.

POEM ROCKET,
ALLEN GINSBERG (1926–97),
American poet

47 Tucanae (NGC 104)

From its perch 16,000 light years away, this glorious globular cluster shines brightly at magnitude 4.5. Although it is a naked eye object under dark conditions, a 4 inch (100 mm) or larger telescope really brings out the best in this cluster, which competes with Omega (φ) Centauri for the title of the most splendid globular cluster in the entire sky. It is more centrally condensed than its rival in Centaurus.

The Small Magellanic Cloud (SMC)

A member of our Local Group, this galaxy is visible to the naked eye on a good night, with the spectacular globular cluster 47 Tucanae alongside. Just a little less than 200,000 light years away, the cloud is some 30,000 light years wide.

TUCANAE (TUC)
On meridian
10 p.m. Oct 20

KEY

x |

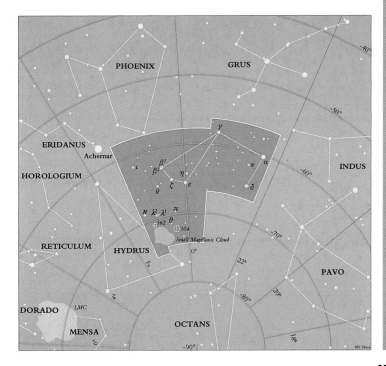

URSA MAJOR (ER-suh MAY-jer)
The Great Bear

This well-known constellation contains the group of seven stars that make up the Big Dipper. In Greek myth, Zeus and Callisto, a mortal, had a son called Arcas. Hera, Zeus's jealous wife, turned Callisto into a bear, and one day, while out hunting, her son, not knowing that the bear was his mother, almost killed her. Zeus rescued Callisto, placing both her and her son, whom he also turned into a bear, in the sky together. Callisto is Ursa Major and Arcas is Ursa Minor.

👁 Mizar (Zeta [ζ] Ursae Majoris) and Alcor
This is the famous apparent double star in the middle of the Big Dipper's handle. The stars are separated by 12 arc minutes and it is thus possible to see them as a pair with the naked eye. Mizar is itself a true binary star, separated by 14 arc seconds.

🔭 M 81
This spiral galaxy can be easily seen through binoculars, even when observing in the city, and it is dramatic when observed under good conditions. The oval disk becomes more apparent with increasing telescope size. M 81 is probably a fair representation of how the Milky Way Galaxy would look from the outside.

🔭 M 82
This is a long, thin peculiar galaxy, just ½ degree from M 81. It appears as a thin, gray nebulosity in a 4 inch (100 mm) telescope, but begins to show some detail in an 8 inch (200 mm) or larger one. Even in large telescopes or photographs, however, it is not clear what type of galaxy this is.

🔭 M 101
This large, spread-out spiral galaxy is visible through small telescopes if the sky is dark enough. It needs a wide field and a low-power eyepiece. At 16 million light years, it is one of the closer spiral galaxies to the Milky Way.

🔭 The Owl Nebula (M 97)
This is an oval planetary nebula that takes the shape of an owl when it is seen in a 12 inch (300 mm) telescope. It is large and dim, and a 3 inch (75 mm) or larger telescope is needed to find it.

STAR OUTBURST IN M 82
M 82 is now thought to be a nearby example of a starburst galaxy. Astronomers attribute the burst of star formation in M 82 to a possible encounter with M 81, its companion spiral galaxy, about 100 million years ago. This may have severely disrupted M 82's gas clouds, igniting the starburst that involved a mass of material equal to several million Suns.

URSAE MAJORIS (UMA)
On meridian
10 p.m. April 20

KEY

x 2.5

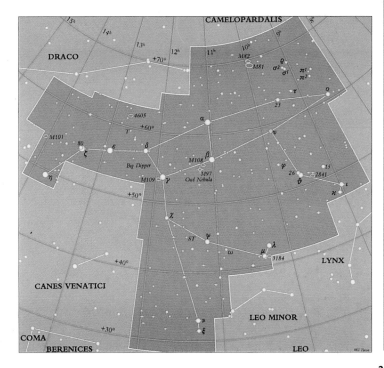

URSA MINOR (ER-suh MY-ner)
The Little Bear

Also known as the Little Dipper, Ursa Minor looks somewhat like a spoon whose handle has been bent back by a playful child. This group of stars was recognized as a constellation in 600 BC by the Greek astronomer Thales. The Little Bear, according to Greek legend, is Arcas, son of Callisto—Ursa Major, the Great Bear. Placed in the heavens by Zeus, he and his mother follow each other endlessly around the north celestial pole.

Polaris (Alpha [α] Ursae Minoris) The pole star for the Northern Hemisphere, this Cepheid variable is currently almost 1 degree from the exact pole. Precession of the Earth's axis will carry the pole to within about 27 arc minutes of Polaris around the year 2100, and then it will start to move away again. Polaris is 820 light years away, with a 9th magnitude companion some 18½ arc seconds away. Splitting this pair is an interesting test for a 3 inch (75 mm) telescope.

KEATS ON POLARIS

While touring England's Lake District in 1819, the poet John Keats, gravely ill with tuberculosis, was thinking of Polaris when he wrote these lines:

> Bright Star, would I were steadfast as thou art—
> Not in lone splendor hung aloft the night
> And watching, with eternal lids apart,
> Like Nature's patient, sleepless Eremite.

URSAE MINORIS (UMI)
On meridian
10 p.m. June 10

KEY

x 1.5

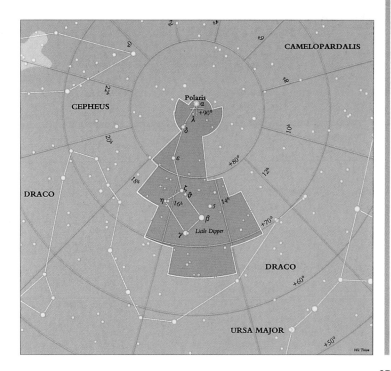

VELA (VEE-lah)
The Sail (of Argo)

This constellation, along with Carina (the Keel), Puppis (the Stern) and Pyxis (the Compass), once formed part of an enormous group of stars in the southern skies known as Argo Navis, the Ship Argo. In Greek myth, this was the vessel that Jason and the Argonauts sailed in on their search for the Golden Fleece. Argo Navis was divided up by Nicolas-Louis de Lacaille in the 1750s, sharing its stars between the four resulting constellations. This left Vela with no stars designated alpha (α) or beta (β).

NGC 3132 This bright planetary nebula accompanies the many clusters in Vela, but lies right on the border with Antlia. Being 8th magnitude and almost 1 arc minute across, it is considered the southern version of Lyra's Ring Nebula, but with a much brighter central star.

The False Cross Delta (δ) and Kappa [κ] Velorum, together with Epsilon (ϵ) and Iota (ι) Carinae, make up a larger but fainter version of the Southern Cross which is known as the False Cross.

Gamma (γ) Velorum This double star is resolvable in a steady pair of binoculars. The primary is a Wolf-Rayet star—very hot and luminous.

VELORUM (VEL)
On meridian
10 p.m. March 10

KEY

x 1.5

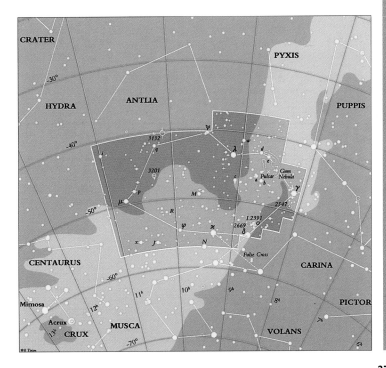

VIRGO (VER-go)
The Maiden, the Virgin

Virgo is the only female figure among the constellations of the zodiac and has been thought to represent a great array of deities since the beginning of recorded history. Among others, she has been identified with the Babylonian fertility goddess Ishtar; Astraea, the Roman goddess of justice; and Demeter, the Greek goddess of the harvest (Roman, Ceres). Virgo is usually shown either holding an ear of wheat or carrying the scales of Libra, the adjoining constellation.

Spica (Alpha [α] Virginis)
A bright white star, Spica is the ear of wheat Virgo is holding. The star is almost exactly 1st magnitude, although it has a slight variation. It is 220 light years away and more than 2,000 times as luminous as the Sun.

Porrima (Gamma [γ] Virginis)
This is one of the best double stars in the sky, each component shining at magnitude 3.7. At 3 arc seconds separation, the pair is still easy to separate, but by the year 2017 they will appear much closer together.

M 84 and 86
These two elliptical galaxies are close enough to be seen in the same low-power telescope field. On a dark night, an 8 inch (200 mm) telescope will show several smaller galaxies in the same view.

M 87
This elliptical galaxy is one of the mightiest galaxies we know. Through a small telescope it appears as a bright patch of fuzzy light about a magnitude brighter than M 84 and 86. Interestingly, larger telescopes don't show a great deal more. In the professional size range, however, more details do emerge. With a 60 inch (1.5 m) telescope, for example, you can see a jet emerging from the galaxy's center.

The Sombrero Galaxy (M 104)
Although this galaxy is quite a distance south of the main concentration of galaxies, it seems to be gravitationally attracted to the swarm and so is thought to be a part of it. The brightest of the Virgo galaxies, a dark lane cuts along its equator, making it look a little like a Sombrero hat in an 8 inch (200 mm) telescope.

REALM OF THE GALAXIES
Scattered throughout Virgo and Coma Berenices are more than 13,000 galaxies. Known as the Virgo Cluster or Coma-Virgo Cluster, this mighty club of distant systems of stars repays sweeping with a small, wide-field telescope on a dark night. To see much detail usually requires an 8 inch (200 mm) or larger telescope.

VIRGINIS (VIR)
On meridian
10 p.m. May 10

KEY

x 3

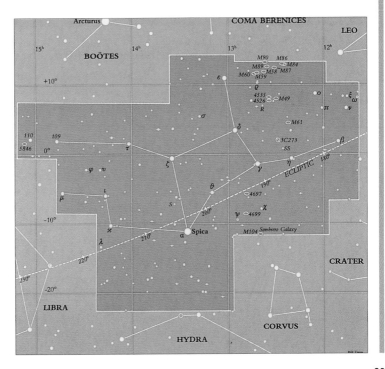

VOLANS (VOH-lanz)
The Flying Fish

The constellation of Piscis Volans, the Flying Fish, lies south of Canopus, and was introduced by Johann Bayer in his Uranometria of 1603. It is now known only as Volans. Sailors in the south seas had reported seeing schools of flying fish, which may have been the inspiration for the name. The pectoral fins of these fish are as large as the wings of birds and they glide across the water for distances of up to $1/4$ mile (400 m).

The stars speak of man's insignificance in the eternity of time; the desert speaks of his insignificance right now.

AUTUMN ACROSS AMERICA,
EDWIN WAY TEALE (1899–1980),
American writer

SKYWATCHING TIP
For long exposure photography of stars and planets, it is important to align the telescope mounting more precisely than for visual observing. This will save you major guiding headaches later. There are several alignment methods, and some companies even manufacture special alignment telescopes that can be attached to your mount's polar axis.

S Volantis A Mira star, S Volantis usually has a maximum magnitude of 8.6, but it has occasionally risen to 7.7. Its faint minimum averages 13.6. The star completes its cycle in a little less than 14 months.

VOLANTIS (VOL)
On meridian
10 p.m. Feb 20

KEY

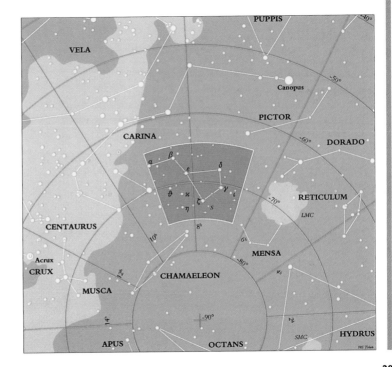

VULPECULA (vul-PECK-you-lah)
The Fox

This constellation, invented by Johannes Hevelius in 1690, is without an exciting story or a moral tale. Hevelius's name for it was Vulpecula cum Anser, the Fox with the Goose, but now the constellation is simply referred to as the Fox.

NEBULAE AND PLANETS
A planetary nebula has nothing to do with planets—it is the last stage of a star's life. It was so-called because when nineteenth-century astronomers viewed it through a telescope, they thought it looked more like the disk of a planet than the point of a star.

 The Dumbbell Nebula (M 27) This is one of the finest planetary nebulae in the sky and is well suited to small telescopes. Bright and large, it is easy to find just north of Gamma (γ) Sagittae. Being 7th magnitude, it can be found through binoculars, but it appears only as a faint nebulous spot. If you use a small telescope, you can make out its odd shape. A larger telescope will reveal its 13th magnitude central star. Although the nebula's gases are expanding at the rate of 17 miles (27 km) per second, there will be no noticeable change in the appearance of the nebula within a human lifetime.

284

VULPECULAE (VUL)
On meridian
10 p.m. Aug 20

KEY

x 2.5

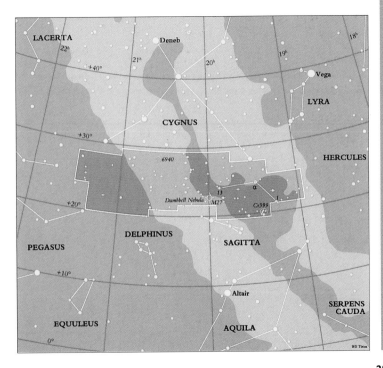

285

Breaking Away
FROM THE
Earth

Perhaps more than any other science, astronomy has an "otherworldly" focus—it looks away from the familiar Earth and gets to grips with the mysterious universe. New tools have helped astronomers unlock some of the secrets of space. Space flight, ever-larger telescopes, more sophisticated detectors for observing the sky—these and other developments have played major roles in the explosive growth in our understanding of the universe that has taken place in the twentieth century.

REACHING OUT

Technological progress in the twentieth century has opened wide our windows on the universe. Today's astronomers have an array of tools at their disposal—electronic cameras, radio telescopes, spacecraft— that their predecessors of just a century ago could not even dream of. But the increasingly sophisticated technology has not obscured the age-old goal: to discover what the universe is and why it is the way it is.

FICTION BECOMES FACT

Jules Verne once imagined what it would be like to travel into space. Today, science fiction has become fact, as this space shuttle astronaut shows. His pressure suit and life-support system keep him alive, while his maneuvring unit allows him to move freely in space.

DESTINATION MARS
The Mars Pathfinder mission of 1997 increased our knowledge of Mars' geology. This image of the Martian surface was taken by the Pathfinder lander.

Astronomy expands In the 1860s, the development of spectroscopy—the splitting of light into its constituent colors—started to shift astronomy from the study of the positions of celestial objects to the study of their physical properties. Astronomers could now classify stars and chart them from birth to death.

Giant telescopes Very large optical telescopes were built in the first half of the twentieth century, and they made significant discoveries. Today, new mirror-making techniques and computer controls are making even grander instruments possible. An entirely new field, radio astronomy, developed after the end of World War II, enabling scientists to measure radio waves emitted by the Milky Way and other galaxies. Astronomers now routinely explore all forms of radiation.

Using space as a base Perhaps the greatest leaps forward have come from astronomical surveys from space, using Earth-orbiting satellites and spacecraft. All-sky surveys from space in the 1970s and 1980s revealed thousands of new objects, including new classes of infrared-bright galaxies, high-energy binary stars and black-hole candidates. The detailed imaging and spectroscopic observations of the Hubble Space Telescope are of incalculable value.

Probing the planets Much of our knowledge of planets has been gained from the observations by spacecraft. These explorations have shown us a Martian volcano dwarfing any on Earth, curious "racetrack" patterns on Uranus's moon Miranda, and possible deposits of ice on the Earth's Moon.

TUNING IN FROM EARTH
The Very Large Array Telescope in New Mexico can resolve great detail from radio signals emanating from deep-space objects.

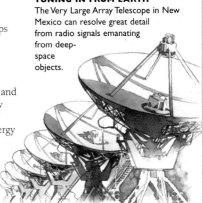

FROM THE GROUND

Our view of space from the ground is improving all the time as telescopes get bigger and more sophisticated, and computers provide new ways to gather and interpret astronomical data.

Seeing better New mirror-making technology has led to the construction of very large telescopes. The 400 inch (10 m) Keck I and Keck II instruments on Mauna Kea, Hawaii, are the world's largest optical scopes. Computers are used to adjust the shape of the telescopes' optics to reduce the distorting effects of Earth's atmosphere and so keep the images sharp. This technique, known as active optics, produces images that rival those taken from space.

CCDs Photographic cameras have now been largely displaced by light-sensitive electronic detectors called charge-coupled devices. CCDs have increased the telescope's light-gathering capability almost a hundredfold and revolutionized imaging. These devices are used in all aspects of astronomical research, from studying variations in starlight to imaging the centers of active galaxies.

Spectroscopy A spectrograph separates white light into its component colors—a spectrum—crossed by numerous dark and bright spectral lines. It then records these lines as a kind of bar code of the object's physical properties. Most of what we know about the chemical composition, temperature and pressure in any astronomical object is encoded in spectral lines.

ANDROMEDA
This image of the Andromeda Galaxy (M 31) is based on observations made by a radio telescope. It shows that the galaxy is rotating. Yellow and red indicate the areas that are rotating away from us, while green and blue indicate areas that are in systematic motion toward us.

Beyond the visible Visible light is radiation that the eye can see. Compared to the total range of the electromagnetic spectrum, visible light constitutes a narrow span. The low-frequency, low-energy part of the spectrum is the domain of radio, millimeter and infrared radiation.

Radio astronomy Radio astronomy is the study of celestial bodies by means of the radio waves that they emit and absorb naturally. It has been invaluable in enabling us to discover the nature and shape of the Milky Way and other galaxies. Most radio telescopes use large metal parabolic dishes, which focus the faint radio signals that are received. When two or more radio telescopes are linked so that their signals are fed to a common receiver, they are jointly termed an interferometer. The greater the distance between the radio telescopes, the finer the detail that can be resolved.

Millimeter and infrared Millimeter-wave telescopes—using radio dishes—are ideal for observing giant molecular clouds in which stars are likely to form. Between the visible and radio domains lies the infrared regime (IR), some of which is accessible from the ground. In the IR, astronomers probe the depths of gas clouds and the center of galaxies using more or less conventional-looking telescopes.

Other wavelengths Radiation of higher energy than visible light needs to be observed from satellites. Conventional telescopes are used to pick up ultraviolet and low-energy X-rays, whereas less refined imaging devices are needed to study sources of high-energy X-rays and gamma rays. This radiation allows astronomers to study very hot objects such as the corona surrounding the Sun. More exotic objects, such as cataclysmic variable systems, are also intense sources of high-energy radiation.

RADIO EYE
Radio astronomy has greatly improved our understanding of star birth, the nature of the Milky Way, the structure of galaxies and the origin of the universe.

STILL WATCHFUL
Although the twentieth century saw the flowering of many different techniques for observing space, optical telescopes, like this one in Hawaii, are still making immensely valuable observations.

THE SPACE RACE

Practical spaceflight grew from the rocket technology of World War II. In the 1950s and 1960s, the Cold War between the Soviet Union and the United States provided a global arena for the competition known as the space race, which began with the launch of Sputnik, the first artificial satellite, in October 1957.

GLENN IN SPACE
Astronaut John Glenn climbs into the cramped capsule of Mercury 6 in preparation for his historic flight into orbit in 1962. Glenn revisited space in 1998 as part of the crew of the space shuttle *Discovery*.

Soviets in front The Soviet Union often upstaged the United States in the early years. The second Sputnik carried a live dog into space. And in April 1961 Yuri Gagarin became the first human in space when he made a single Earth orbit aboard Vostok 1.

U.S. fightback Three weeks after Gagarin's feat, Alan Shepard in a U.S. Mercury spacecraft traveled beyond Earth's atmosphere but did not go into orbit. John Glenn was the first American to reach orbit, in February 1962; his Mercury 6 craft made a total of three orbits.

More Soviet firsts In June 1963 Valentina Tereshkova became the first woman in space. October 1964 saw three cosmonauts in Voskhold 1 orbit the Earth, five months before the first two-man Gemini mission. The first spacewalk was made by Aleksei Leonov in March 1965, three months before Edward White spacewalked from Gemini 4.

Lunar goal The U.S. program by now had the explicit goal of reaching the Moon. Gemini flights perfected techniques for the rendezvous and docking of craft in orbit—the first docking maneuver was completed by Gemini 8 in March 1966. These techniques were important for lunar trips, which would use spacecraft modules that could be linked in orbit and discarded when no longer needed. Not until 1969 did the Soviets achieve a succcessful docking and crew transfer. By then, however, the Americans were orbiting astronauts around the Moon.

"WE HAVE LIFT-OFF" The Apollo astronauts aboard this Saturn V rocket are heading for a rendezvous with the Moon. Apollo was the culmination of the U.S. space program of the 1960s.

MAN ON THE MOON

In May 1961, President Kennedy committed the United States to landing a man on the Moon by 1970. NASA (National Aeronautics and Space Administration) set out to achieve that goal. Its Mercury and Gemini programs paved the way for Apollo, which, in just six short years, changed the boundaries of human aspiration.

A Lunar Roving Vehicle, or Moon buggy, was first taken to the Moon by Apollo 15 in 1971.

A tragic start Apollo had a disastrous beginning. On January 27, 1967, an electrical fire swept through an Apollo spacecraft on the ground, killing all three astronauts on board. The program was grounded for more than a year until modifications made the spacecraft safer. Apollos 4, 5 and 6 were unpiloted flights designed to test the new systems and refine procedures for putting hardware into orbit. Their success led to the piloted Apollo 7 mission in October 1968, and the first lunar orbit by humans in December 1968.

LONG-LASTING PRINTS
Twelve men have stepped onto the lunar surface. Their footprints will remain there for many millions of years.

Apollos 9 and 10 Apollo 9 tested the lunar module while orbiting the Earth in March 1969. Carrying two of the three crew, this spindly legged craft would separate from the command module and descend to the lunar surface. After spending several hours on the Moon, the two astronauts would lift off in the upper part of the lander and rendezvous with the command module for the journey back to

Earth. In May 1969 the crew of Apollo 10 flew a full dress rehearsal: they traveled to the Moon, deployed the lunar module, and took it to within 50,000 feet (15,000 m) of the landing site, the Sea of Tranquility (Mare Tranquillitatis). The lunar surface was now within reach.

One giant leap Apollo 11, with Neil Armstrong, Edwin "Buzz" Aldrin and Michael Collins aboard, lifted off on July 16, 1969. Reaching lunar orbit on July 20, Armstrong and Aldrin climbed into the lunar module, the *Eagle*, and began their descent. Soon after, the *Eagle* touched down on a flat area of the Sea of Tranquility. Armstrong stepped onto the Moon and uttered, "That's one small step for a man, one giant leap for mankind." Aldrin joined him, and the pair spent more than two hours on the surface, placing experiments and collecting rocks. The crew's return to Earth was a political and scientific triumph.

Later landings Flights to the Moon continued. Apollo 12 stayed longer and collected more samples. Then came the drama of Apollo 13. An oxygen tank exploded en route to the Moon, putting the astronauts in extreme danger. Only heroic efforts and ingenious improvization saved their lives. Apollo 14 restored American confidence in lunar flight, and the last three missions—Apollos 15, 16 and 17—targeted areas with spectacular terrain. In December 1972 the Apollo program slipped into history. The task of venturing to other worlds has been largely turned over to unmanned probes.

Apollo's achievements The six Apollo missions returned 844 pounds (382 kg) of rock and soil samples, amazing photographs of lunar features, and a vast quantity of data generated by instruments set up on the surface. Quite simply, Apollo's discoveries laid the basis for our modern understanding of the Moon.

APOLLO 11
"Buzz" Aldrin steps onto the lunar surface as part of the 1969 mission. He and Neil Armstrong spent more than two hours walking on the Sea of Tranquility.

THE OUTER LIMITS

In 1930, Clyde Tombaugh enlarged our Solar System with his discovery of Pluto. The outer limits of the universe have been pushed back continuously since then, with new discoveries coming thick and fast. In fact, one of the most exciting things about the astronomy described in this book is how new it all is. Quite literally, much of it could not have been written until the last few years. For example, we now know of a dozen or so planetary systems besides our own. And space probes have allowed us to get to know our Solar System far more intimately. This process of exploration will go on pushing back the limits of our knowledge—a long way beyond Tombaugh's new planet.

SPACE SHUTTLES AND STATIONS

In the early 1970s, NASA started work on the first reusable spaceship, the space shuttle, which could carry up to seven astronauts and a cargo of more than 30 tons (27 tonnes). At about the same time, the Soviet Union launched the first manned space station, Salyut. Shuttles and stations promise to be the future of manned space missions.

MEN AT WORK
Two astronauts ready a communications satellite for launch from the payload area of a space shuttle. Space walks like these have become routine.

INTERNATIONAL SPACE STATION
Largely an initiative of the United States, the station will use components from 14 other countries, including Russia, Japan, France, Germany and Britain.

Columbia flies The first space shuttle, *Columbia*, was launched on April 12, 1981, exactly 20 years after Yuri Gagarin's flight in Vostok 1. Three more shuttles soon joined the fleet, *Challenger, Discovery* and *Atlantis*. A catastrophic explosion destroyed *Challenger* in 1986; *Endeavor* replaced it.

Shuttle basics The shuttle has three components: the orbiter, the external tank and the solid rocket boosters. The orbiter has a payload bay large enought to hold a passenger bus. The external tank

fuels the engines during launch and burns up in the atmosphere, while the boosters that help propel the shuttle off Earth's surface are ejected prior to orbit. These parachute into the Atlantic Ocean, where they are retrieved and refurbished. Once its mission in space is accomplished, the shuttle re-enters Earth's atmosphere and glides to a landing.

Space truck The shuttle is an all-purpose "space truck." It delivers satellites into orbit and retrieves them for repair, and can serve as the first stage for probes to other planets. Reusable space laboratories have flown aboard the shuttle many times, conducting scientific experiments and astronomical observations.

Salyut leads the way The Soviet Union launched seven space stations, all named Salyut, between 1971 and 1982. Crew were transported to and from the station by a Soyuz craft. Over the course of the Salyut program, the Soviets pioneered the study of long-term weightlessness. The flight plan was for a crew to spend up to a year in the station, while other crews (including some

A JOURNEY STARTS
A space shuttle is launched into the skies above Cape Canaveral. Each of the shuttle's main engines has enough thrust to power two and a half jumbo jets. In contrast to this explosive beginning, at the end of its mission the shuttle will glide, unpowered, to the ground.

non-Russians) arrived for short visits. Besides conducting medical and psychological tests, the crews made astronomical observations and studied Earth.

Skylab The United States launched Skylab in May 1973. Hundreds of experiments and observations were made by three crews of astronauts.

Mir Following the Salyut series came a more ambitious craft named Mir ("peace"), which has been occupied almost continuously since its launch in February 1986. It has been visited by astronauts from a number of countries in a prelude to the construction of the International Space Station.

Under construction The International Space Station is being built with hardware and ideas from many countries. Measuring 356 by 290 feet (110 by 90 m), the station will house up to seven crew, who will conduct a range of scientific experiments. The station will also demonstrate the effects of extended space travel on humans, which should help in the planning of manned expeditions to the Moon and Mars.

THE HUBBLE SPACE TELESCOPE

Earth's atmosphere distorts and filters the cosmic radiation passing through it. To overcome this problem, scientists have positioned telescopes above the atmosphere by placing them in rockets and satellites. Most famous of these is the Hubble Space Telescope (HST), which has made astonishing discoveries.

What Hubble does Launched in 1990, HST observes and provides images and spectroscopy in visual, near-infrared and ultraviolet wavelengths of stars, the thin interstellar gas and galaxies. Consisting of a 95 inch (2.4 m) mirror and a suite of sensitive instruments, it is controlled by 400 astronomers, computer scientists and technicians.

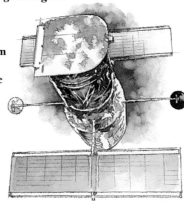

HUBBLE IN SPACE
The Hubble Space Telescope, named for American astronomer-cosmologist Edwin Hubble, has produced images far sharper than any taken from Earth.

Sensitive eyes The Wide Field/Planetary Camera II, the most often used of HST's instruments, can detect objects as faint as 28th magnitude (about a billion times fainter than can be seen with the naked eye). The Faint Object Camera can also record 28th magnitude objects, but it offers higher resolution and a wider choice of viewing angles.

Other instruments Two other instruments are the Near Infrared Camera and Multi-Object Spectrometer (NICMOS), and the Space Imaging Spectrograph (STIS). NICMOS handles both imaging and spectroscopic observations of objects at near-infrared wavelengths. It aims to tells us much about the birth of stars in dense, dusty globules, the infrared emission produced by the active centers of distant galaxies, and the nature of a class of galaxies as bright as quasars at infrared wavelengths. The STIS covers a broad range of wavelengths and can also block out, or occult, the light of distant stars to search for black holes. Finally, HST's fine-guidance sensors, necessary for pointing the telescope and locking onto its target, can measure the positions of stars to 0.002 arc seconds.

What HST has seen By the late 1990s, HST had looked at well over 10,000 objects and made more than 100,000 exposures; yielded significant insights into the formation of stars and stellar disks; disclosed important evidence for the existence of black holes in galaxies and quasars; increased our knowledge of the size and age of the universe; and detected galaxies that formed only a billion years after the Big Bang. Its high-resolution images of Mars, Jupiter, Saturn and Neptune are yielding details surpassed only by space-probe photographs.

SATURN'S AURORA
HST took this ultraviolet image of Saturn's auroral cloud formations and rings. These images are yielding details surpassed only by space-probe photos.

The stars close-up The detail of HST's observations is so great that astronomers can now see, for example, great chunks of matter swirling around supermassive black holes at the centers of galaxies and quasars, as well as structural details in the spiral arms of nearby galaxies.

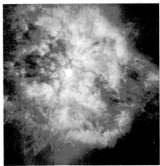

COSMIC FIREWORKS
Hubble's image of the energetic star WR124 reveals it is surrounded by hot clumps of gas being ejected into space. The star is 15,000 light years away.

The future Plans are already afoot to replace HST with a more powerful space telescope. This will be able to look in even greater detail at a period in the universe when the promordial seeds of the galaxies began to evolve. Study of this epoch may help explain the origin and fate of the universe.

SPACE PROBES

From the 1960s, a succession of spacecraft began to explore the planets. Controlled from Earth by radio, and reporting back in a stream of digital data, these robots have changed our view of the Solar System's other worlds from enigmatic dots of light in a telescope to places as real as Earth.

MARS MISSION
Mars Global Surveyor left Earth in 1996, the first in a series of spacecraft that will study the Red Planet in detail.

THE SUN BY SOHO
The SOHO probe returned this ultraviolet image of the Sun in 1997. Scientists hope SOHO will make observations until 2003.

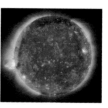

MAGELLAN'S LAUNCH
Magellan is seen here being launched from the space shuttle Atlantis in 1989. Arriving at Venus in 1990, the probe used radar to peer through the planet's clouds, and provided an astonishingly detailed view of the topography of the planet. Armed with this information, scientists have begun to unravel Venusian history. Contact with Magellan was lost in 1994.

VOYAGER

Voyager 2 left Earth in 1977 to fly past all four gas-giant planets. Among many other achievements, it made a close approach to Neptune, revealing an aqua-blue atmosphere.

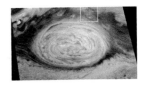

GALILEO

Galileo recorded this image of Jupiter's Great Red Spot. It also deployed a probe into the planet's atmosphere.

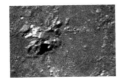

Milestones The first probe to reach another planet was Mariner 2, which flew past Venus in 1962. Other milestones include:
• 1971–72: first detailed close-range pictures of Mars (Mariner 9)
• 1973: first fly-by of Jupiter (Pioneer 10)
• 1974: first close-range pictures of Mercury and Venus cloud tops (Mariner 10)
• 1975: first images from the surface of Venus (Venera 9)

CLEMENTINE

In 1994 Clementine surveyed the Moon's topography and composition from pole to pole. Shown here is the central peak of the crater Tycho. The probe found what may be ice deposits at the south pole.

• 1976: Vikings 1 and 2 land on Mars and begin observations of terrain and climate
• 1979: two fly-by missions to Jupiter (Voyagers 1 and 2); first fly-by of Saturn (Pioneer 11)

• 1980: first detailed study of Saturn and its system (Voyager 1)
• 1981: second fly-by of Saturn (Voyager 2)
• 1986: Voyager 2 fly-by of Uranus; Giotto intercepts comet Halley.
• 1989: Voyager 2 fly-by of Neptune
• 1994: Magellan completes radar mapping of Venus
• 1995: Galileo deploys probe into Jupiter's atmosphere
• 1997: geological survey of Mars by Mars Pathfinder and Global Surveyor; launch of the Cassini–Huygens mission to Saturn (due to arrive in 2004)

OTHER SOLAR SYSTEMS

Throughout history, the only planets that anyone knew about were the Earth and its sisters orbiting the Sun. Then, in 1995, a discovery sent shock waves around the world. For the very first time, a planet had been detected outside our Solar System. And that was only the beginning . . .

First discovery The discovery of a planet orbiting a star called 51 Pegasi in the constellation Pegasus opened the floodgates. More than a dozen planets have now been recognized, with many more tentative findings.

How to find a planet The extra-solar planets are so faint compared to their parent star that they cannot be seen directly. So astronomers find them by looking for the tiny wobbles the planets' gravity induces in the motion of each star.

Up close The new planets are very large and orbit very close to their stars. Astronomers believe that some-thing—perhaps interactions with the disk from which the planets first formed—may have dragged these giant bodies close to their star. So far, no Earth-size planets have been found, but the search continues.

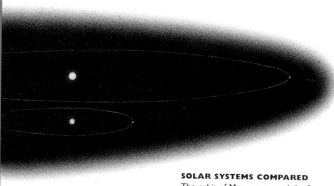

SOLAR SYSTEMS COMPARED
The orbit of Mercury around the Sun (top) is shown on a similar scale to Rho Cancri and its planet (bottom).

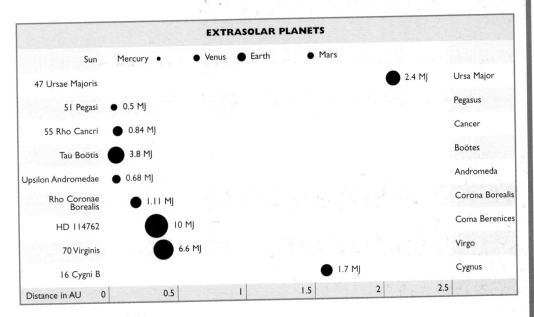

EXTRASOLAR PLANETS

Sun   Mercury •   ● Venus ● Earth   ● Mars

47 Ursae Majoris   ● 2.4 MJ   Ursa Major

51 Pegasi   ● 0.5 MJ   Pegasus

55 Rho Cancri   ● 0.84 MJ   Cancer

Tau Boötis   ● 3.8 MJ   Boötes

Upsilon Andromedae   ● 0.68 MJ   Andromeda

Rho Coronae Borealis   ● 1.11 MJ   Corona Borealis

HD 114762   ● 10 MJ   Coma Berenices

70 Virginis   ● 6.6 MJ   Virgo

16 Cygni B   ● 1.7 MJ   Cygnus

Distance in AU   0   0.5   1   1.5   2   2.5

NEW AND MASSIVE WORLDS

In this chart, nine of the new planets are compared with the inner planets of our Solar System (distances are shown to scale). The figures beside each planet compare its mass with the mass of Jupiter (MJ). Most of these massive planets orbit remarkably close to their stars. The sizes of the new planets are still to be determined.

THE FUTURE IN SPACE

With ever more sophisticated technology and mountains of data pouring in, it is difficult to imagine what revelations even the next 20 years of space exploration will bring. Nevertheless, here are a few predictions.

A SETTLEMENT ON MARS

Will humans ever live on other worlds? This artist's conception of a human settlement on Mars may seem far-fetched, but spacecraft such as the space shuttle and X-34 also seemed far-fetched not long ago. The decades ahead promise a much deeper understanding of our Solar System and of Mars in particular.

Future missions

Tight budgets have discouraged large, expensive space-probe missions. The emphasis now is on flying smaller spacecraft more often. The following missions are planned for the early years of the twenty-first century:

• The European Space Agency (ESA) plans to send SMART 1, an orbiter, to the Moon in 2001.

• In 2001, NASA's Genesis mission will collect samples from the solar wind and return them for analysis, giving scientists data to test theories of how the Sun and planets formed.

PLUTO AHEAD

The Pluto Kuiper Express will be the first spacecraft to visit Pluto and its moon Charon. After studying the Pluto–Charon system, it will travel to the Kuiper Belt.

• NASA's CONTOUR probe, to be launched in 2002, will photograph three comets at close range.

• ESA's Rosetta spacecraft is to be launched in 2003. It will rendezvous with comet Wirtanen in 2011, and deploy a lander to its surface.

• Japan's Institute for Space and Astronautics (ISAS) will launch a lunar orbiter–lander in 2003.

• NASA plans to launch Pluto Kuiper Express in 2004. It should arrive at Pluto in 2012, then continue into the Kuiper Belt, from where it will leave the Solar System.

• The Cassini–Huygens mission—launched in 1997—is due to arrive at Saturn in 2004. The Huygens probe will be deployed to explore the surface of the moon Titan.

• Mercury will be studied by two probes—ISAS's orbiter is to be launched in 2005, while ESA's craft will follow four years later.

Mars Scientists have particularly ambitious plans for Mars. ESA's InterMarsNet is an orbiter with a collection of landers designed to study the planet's internal structure. Arriving in 2004, the landers would operate for one Martian year. Mars Global Surveyor, which left Earth in 1996, was the first in a planned decade-long exploration of Mars by NASA, involving a succession of orbiters, landers and rovers.

The X-planes NASA is developing the X-34, a single-engined rocket-powered "space plane" that will fly itself using onboard computers. It will travel eight times faster than the speed of sound. Carried aloft by an airliner, it fires its engines after separation from the mother plane. Work is also taking place on the X-38, which will be the emergency "lifeboat" for crew of the International Space Station. It may also have other functions, such as bringing cargo to the space station.

X-38
The X-38 emergency crew return vehicle is designed for use as a "lifeboat" for the crew of the International Space Station, now under construction. The craft should begin operations aboard the station in 2003.

Glossary

accretion disk A flat sheet of gas and dust surrounding a newborn star, a black hole or any massive object growing in size by attracting material.

active galaxy A galaxy with a central black hole emitting lots of radiation.

apparent magnitude The brightness of a star (or any celestial object) as seen from Earth.

astronomical unit (AU) The average distance between Earth and the Sun—about 93 million miles (150 million km).

atmosphere The layer of gases enveloping a celestial object.

axis The imaginary line through the center of a planet, star or galaxy around which it rotates.

Big Bang Cosmologists' best theory for the origin of the universe: that it began as an explosion of a tiny, superhot bundle of matter some 15 billion years ago.

binary star (double star) Two stars linked by mutual gravity and revolving around a common center of mass.

black hole A massive object so dense that no light or other radiation can escape from it.

CCD (charge-couple device) A computer-controlled electronic detector that can record an image.

celestial equator The imaginary line encircling the sky midway between the two celestial poles.

celestial poles The imaginary points on the sky where Earth's rotation axis, extended infinitely, would touch the imaginary celestial sphere.

celestial sphere The imaginary sphere enveloping the Earth upon which the stars, galaxies and other celestial objects all appear to lie.

circumpolar stars Stars located near either the celestial north or south pole and never setting when viewed from a given location.

constellation One of the 88 official patterns of stars into which the night sky is divided.

declination The angular distance of a celestial object north or south of the celestial equator; corresponds to latitude.

dwarf star A star, such as the Sun, that lies on the main sequence.

ellipse The oval, closed path followed by a celestial object moving under gravity; for example, a planet around the Sun.

fireball Any meteor brighter than Venus, about magnitude −4.

galactic (or open) star cluster A group of some few hundred stars bound by gravity and moving through space together.

galaxy A huge gathering of stars, gas and dust, bound by gravity and having a mass ranging from 100,000 to 10 trillion times that of the Sun. There are spiral, elliptical and irregular types of galaxies.

globular star cluster A spherical cluster that may contain up to a million stars.

Hertzsprung-Russell (HR) Diagram A graph whose horizontal axis plots star color (or temperature) against a vertical axis plotting stellar luminosity (or absolute magnitude).

light year The distance that light travels in one year: 6 trillion miles (10 trillion km)

luminosity The total intrinsic brightness of a star or galaxy.

magnitude A logarithmic unit used to measure the optical brightness of celestial objects. Numerically lower magnitudes are brighter than numerically larger magnitudes. A five-magnitude difference represents a 100-fold change in brightness.

main sequence The band on the Hertzsprung-Russell Diagram where stars lie for much of their life.

meridian An imaginary line on the sky that runs due north and south and passes through your zenith.

meteor A piece of space debris which produces a bright, transient streak of light by burning up as it enters the atmosphere at high speed.

meteorite Any piece of interplanetary debris that reaches the Earth's surface intact.

nebula Any cloud of gas or dust in space. May be luminous or dark.

neutron star A massive star's collapsed remnant that consists almost wholly of neutrons. May be visible as a pulsar.

nova A white dwarf star in a binary system that brightens suddenly by several magnitudes as gas pulled away from its companion star explodes.

parallax The apparent change in position of a nearby star due to Earth's orbital motion around the Sun.

parsec A unit of distance equal to 3.26 light years. It is the distance at which a star would have a parallax of 1 second of arc.

precession A slow periodic wobble in the Earth's axis caused by the gravitational pull of the Sun and Moon.

pulsar A rapidly spinning neutron star that flashes periodic bursts of radio (and, in some cases, higher) energy.

quasar A compact, quasi-stellar object or radio source, the spectrum of which displays a marked redshift. Thought to be the active cores of very distant galaxies.

red giant A large reddish or orange star in a late stage of its evolution. It is relatively cool and has a diameter perhaps 100 times its original size.

redshift An apparent shift of spectral lines in the light of an object toward the

red, caused by relative motion between the object and Earth.

retrograde motion An apparent westward motion of a planet, asteroid or comet relative to the stars.

right ascension (RA) The celestial coordinate analogous to longitude on Earth.

sidereal period The time, *relative to the stars*, needed for a planet or moon to make one rotation or revolution around its primary body.

supernova The explosion of a massive star in which it blows off its outer layers of atmosphere and briefly equals a galaxy in brightness.

supernova remnant The gaseous debris thrown off by a supernova.

variable star Any star, the brightness of which appears to change, with periods ranging from minutes to years.

white dwarf The small, very hot remnant of a star that evolved past the red giant stage.

zenith The point on the celestial sphere directly overhead.

INDEX

Page numbers in *italics* indicate illustrations and photos.

ACKNOWLEDGMENTS

TEXT CREDITS

Robert Burnham, Alan Dyer, Robert A. Garfinkle, Martin George, Jeff Kanipe, David H. Levy, Dr John O'Byrne, Jo Rudd (index), Gabrielle Walker.

PICTURE CREDITS

Ad-Libitum (Weldon Owen Pty Ltd), Astro Photo/Tony & Daphne Hallas, Corel Corporation, NASA, Stockbyte.

ILLUSTRATION CREDITS

Steven Bray, Gregory Bridges, John Bull, Lynette R. Cook, Nick Farmer/Brihton Illustration, Robert Hynes, Mike Lamble, Rob Mancini, Peter Mennim, Trevor Ruth, Darren Pattenden/illustration, Oliver Rennert, Ray Sim, Kevin Stead, Oliver Strewe, Steve Travaskis, Wil Tirion (star maps), Genevieve Wallace, David Wood.